T0350133

THE CHOROID PLEXUS
AND CEREBROSPINAL FLUID

ELSEVIER *science &
technology books*

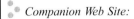

Companion Web Site:

http://store.elsevier.com/product.jsp?&isbn=9780128017401

The Choroid Plexus and Cerebrospinal Fluid
Josh Neman and Thomas C. Chen

Resources:

• Abstracts

• Tables and figures (PowerPoint presentation)

• References

• Glossary

ELSEVIER

ACADEMIC
PRESS

THE CHOROID PLEXUS AND CEREBROSPINAL FLUID

Emerging Roles in CNS Development, Maintenance, and Disease Progression

Edited by

JOSH NEMAN
Department of Neurosurgery, Keck School of Medicine,
University of Southern California, Los Angeles, CA, USA

THOMAS C. CHEN
Department of Neurological Surgery, Keck School of Medicine,
University of Southern California, Los Angeles, CA, USA

ELSEVIER

AMSTERDAM • BOSTON • HEIDELBERG
LONDON • NEW YORK • OXFORD • PARIS
SAN DIEGO • SAN FRANCISCO • SINGAPORE
SYDNEY • TOKYO
Academic Press is an imprint of Elsevier

Academic Press is an imprint of Elsevier
125, London Wall, EC2Y 5AS, UK
525 B Street, Suite 1800, San Diego, CA 92101-4495, USA
225 Wyman Street, Waltham, MA 02451, USA
The Boulevard, Langford Lane, Kidlington, Oxford OX5 1GB, UK

British Library Cataloguing-in-Publication Data
A catalogue record for this book is available from the British Library

Library of Congress Cataloging-in-Publication Data
A catalog record for this book is available from the Library of Congress

ISBN: 978-0-12-801740-1

For information on all Academic Press publications
visit our website at http://store.elsevier.com/

Typeset by Thomson Digital

Publisher: Mica Haley
Acquisition Editor: Natalie Farra
Editorial Project Manager: Kristi Anderson
Production Project Manager: Lucía Pérez
Designer: Matthew Limbert

Working together
to grow libraries in
developing countries

www.elsevier.com • www.bookaid.org

Dedication

To my wife, Marjan, the one true constant and enduring supporter of my scientific journey.

– Josh Neman

To my mother, Li Hua Chen, who always believed in me.

– Thomas C. Chen

Contents

1. Introduction to the Ventricular System and Choroid Plexus

TATSUHIRO FUJII, JOSHUA YOUSSEFZADEH, MICHAEL NOVEL, JOSH NEMAN

2. Development of Brain Ventricles and Choroid Plexus

ELLEN M. CARPENTER

3. Choroid Plexus: Structure and Function

FLORENCE M. HOFMAN, THOMAS C. CHEN

List of Contributors

Jérôme Badaut UMR 5287-Institut de Neurosciences Cognitives et Intégratives d'Aquitaine, Université de Bordeaux, Bordeaux, France

Thomas Brinker Department of Neurosurgery, Warren Alpert Medical School, Brown University, Rhode Island Hospital, Providence, RI, USA

Ellen M. Carpenter Department of Psychiatry and Biobehavioral Sciences and Interdepartmental Program in Neuroscience, David Geffen School of Medicine, University of California, Los Angeles, CA, USA

Marc C. Chamberlain Department of Neurology and Neurological Surgery, Seattle Cancer Care Alliance, Fred Hutchinson Cancer Research Center, University of Washington, Seattle, WA, USA

Thomas C. Chen Department of Neurological Surgery, Keck School of Medicine, University of Southern California, Los Angeles, CA, USA

Cecilia Choy Division of Neurosurgery, City of Hope and Beckman Research Institute, Duarte, CA, USA

Tatsuhiro Fujii Department of Neurosurgery, Keck School of Medicine, University of Southern California, Los Angeles, CA, USA

Jean-François Ghersi-Egea Blood–Brain Interface Group, Oncoflam Team, and BIP Platform, INSERM U1028, CNRS UMR5292, Lyon Neuroscience Research Center, Faculté de Médecine RTH Laennec, Lyon, France

Sean A. Grimm Cadence Health Brain and Spine Tumor Center, Warrenville, IL, USA

Florence M. Hofman Department of Pathology, Keck School of Medicine, University of Southern California, Los Angeles, CA, USA; Department of Neurological Surgery, Keck School of Medicine, University of Southern California, Los Angeles, CA, USA

Alex Julian Department of Neurological Surgery, Keck School of Medicine, University of Southern California, Los Angeles, CA, USA

Vahan Martirosian Department of Neurological Surgery, Keck School of Medicine, University of Southern California, Los Angeles, CA, USA

John Morrison   Department of Neurosurgery, Warren Alpert Medical School, Brown University, Rhode Island Hospital, Providence, RI, USA

Josh Neman   Department of Neurosurgery, Keck School of Medicine, University of Southern California, Los Angeles, CA, USA

Michael Novel   Department of History, University of California at Los Angeles, Los Angeles, CA, USA

Joshua Youssefzadeh   Department of Neurosurgery, Keck School of Medicine, University of Southern California, Los Angeles, CA, USA; Department of Health Promotion and Disease Prevention, Keck School of Medicine of USC, Los Angeles, CA, USA

About the Editors

JOSH NEMAN

University of Southern California, CA, USA

Dr Neman is a research Assistant Professor of Neurosurgery and member of the Norris Comprehensive Cancer Center at the Keck School of Medicine of the University of Southern California. Dr Neman received his PhD in Neurobiology at the UCLA David Geffen School of Medicine. He then went on to complete his postdoctoral fellowship in Cancer Biology at the City of Hope's Beckman Research Institute. Dr Neman's current research investigates the bi-ology of primary and metastatic brain tumors. His expertise and strengths in neuroscience, cancer, and stem biology have allowed for the development of novel approaches to study the brain and tumor microenvironment, a vantage point that is currently lacking in the field of primary and metastatic brain tumors. Dr Neman has published in PNAS, Cancer Research, PLoS One, Spine, Neurosurgery, and Developmental Neurobiology, and has numerous reviews, abstracts, and book chapters to his name. He is coeditor of the second edition of "Case Files Neuroscience" in McGraw Hill Medical's LANGE Case Files series, which was published in October 2014. Dr Neman has been the recipient of multiple research awards including those from National Institutes of Health/National Cancer Institute, California Institute for Regenerative Medicine (CIRM), American Cancer Society, and Susan G Komen Breast Cancer Foundation.

THOMAS C. CHEN

University of Southern California, CA, USA

Dr Chen is currently Professor of Neurosurgery and Pathology at the Keck School of Medicine of the University of Southern California. Dr Chen is a physician, a board certified neurosurgeon, and the Director of Surgical Neuro-oncology, recognized for his skills as a neurosurgeon and his cutting edge research examining glioma biology. Dr Chen is the head of the Glioma Research Group at USC where he focuses on the area of translational research aimed at the development of clinical trials and novel therapeutics for malignant brain tumors. He received his MD from the University of California, San Francisco before completing his neurological surgery residency and PhD in pathobiology at the University of Southern California. He subsequently completed his fellowship training in spinal surgery from the Medical College of Wisconsin.

Preface

Cerebrospinal fluid (CSF) functions to bathe the brain and spinal cord. It is vital for the maintenance of fluid homeostasis within the central nervous system (CNS). The production of CSF by the choroid plexus is tightly regulated providing the necessary nutrients and removing waste that may compromise the normal homeostasis of the CNS.

Yet the choroid plexus and CSF have been one of the most understudied tissues in neuroscience. However, recent work has begun to elucidate how the choroid plexus and CSF regulate development and disease in ways that extend far beyond traditional neurobiology roles.

As editors, it is our hope to combine new and established work to allow crossdisciplinary discussion from neurosciences (clinical and basic sciences), immunology, and cancer biology to showcase newfound excitement surrounding the choroid plexus and CSF.

Josh Neman
Thomas C. Chen

Acknowledgment

We would like to thank the Elsevier team that helped bring this project to fruition: Natalie Farra and Kristi Anderson.

Josh Neman and Thomas C. Chen

List of Abbreviations

AB	Amyloid beta
ABC	ATP binding cassette
AChA	Anterior choroidal arteries
AD	Alzheimer's disease
AICA	Anterior inferior cerebellar artery
ALS	Amylotrophic lateral sclerosis
ANP	Atrial natriuretic peptide
APC	Antigen presenting cells
APP	Amyloid precursor protein
AQP4	Aquaporin 4
AT/RT	Atypical teratoid/rhabdoid tumors
AVP	Arginine vasopressin
AZT	Azidothymidine
BBB	Blood–brain barrier
BCPB	Blood–choroid plexus barrier
BDNF	Brain-derived neurotrophic factor
BLMB	Blood–leptomeninges barrier
BMP	Bone morphogenic protein
BMP4	Bone morphogenetic protein 4
Brdu	5-Bromo-2-deoxyuridine
CAM	Cell adhesion molecule
cAMP	Cyclic AMP
ChAT	Choline acetyltransferase
CNS	Central nervous system
COX2	Cyclooxygenase 2
CP	Choroid plexus
CPC	Choroid plexus carcinoma
CPE	Choroid plexus epithelial cell
CPP	Choroid plexus papilloma
CPTs	Choroid plexus tumors
CSF	Cerebrospinal fluid
CSI	Craniospinal axis radiotherapy
CT	Computed tomography
CTCs	Circulating tumor cells
DC	Dendritic cells
ECGR-4	Esophageal cancer-related gene-4
ECM	Extracellular matrix
ECS	Extracellular space
eCSF	Embryonic cerebrospinal fluid
EGF	Epidermal growth factor
EGFR	Epidermal growth factor receptor
EMA	Epithelial membrane antigen
GLP-1	Glucagon-like peptide-1
GM-CSF	Granulocyte-macrophage colony-stimulating factor
GTR	Gross total resection

HGF	Hepatocyte growth factor
hNGF	Human neuronal growth factor
HRP	Horse radish peroxide
ICA	Internal carotid artery
ICV	Intraventricular or intracisternal
IGF-I	Insulin-like growth factor-I
IGF-II	Insulin-like growth factor-II
ISF	Interstitial fluid
JAM	Junctional adhesion molecule
L1CAM	L1 cell adhesion molecule
LDL	Low-density lipoprotein
LPS	Lipopolysaccharides
MBP	Metronomic biofeedback pump
MCI	Mild cognitive impairment
MDR	Multidrug resistance protein
MMP	Matrix metalloproteinase
MRI	Magnetic resonance imaging
MRS	Magnetic resonance spectroscopy
MSCs	Mesenchymal stem cells
NAA	N-Acetylaspartate
PCA	Posterior cerebral artery
PChA	Posterior choroidal arteries
PCoA	Posterior communicating artery
PECAM-1	Platelet endothelial cell adhesion molecule-1
PET	Positron emission tomography
P-GP	P-glycoprotein
PI3K	Phosphoinositide-3-kinase
PiB	Pittsburgh compound B
PICA	Posterior inferior cerebellar artery
PSGL-1	P-selectin glycoprotein ligand-1
PVS	Perivascular spaces
RAGE	Receptor for advanced glycation end products
rCBV	Cerebral blood volume
ROCKs	Rho-associated kinases
ROS	Reactive oxidative stress
RT	Radiation therapy
SC	Spinal cord
SCA	Superior cerebellar artery
SDF-1a	Stromal cell-derived factor 1-alpha
Shh	Sonic hedgehog
STR	Subtotal resection
TEM	Transendothelial migration
TGFB	Transforming growth factor beta
TGFb1	Transforming growth factor b1
TLR4	Toll-like receptor 4
TNF	Tumor necrosis factor
TTR	Transthyretin
VEGF	Vascular endothelial growth factor
VP	Vasopressin
WBRT	Whole brain radiation therapy

1

Introduction to the Ventricular System and Choroid Plexus

Tatsuhiro Fujii*, Joshua Youssefzadeh*,†,
Michael Novel**, Josh Neman*

*Department of Neurosurgery, Keck School of Medicine, University of Southern California, Los Angeles, CA, USA; †Department of Health Promotion and Disease Prevention, Keck School of Medicine of USC, Los Angeles, CA, USA; **Department of History, University of California at Los Angeles, Los Angeles, CA, USA

DEVELOPMENT OF THE VENTRICULAR SYSTEM

Within the first 4 weeks of human development, the formation of the central nervous system (CNS) has begun to take shape in the form of the neural tube. From within this enclosed cavity emerge the future ventricles of the brain as well as the central canal of the spinal cord. As the primitive neural tube continues to enlarge in way of rapid cell division, the appearance of the pontine flexure

The Choroid Plexus and Cerebrospinal Fluid. http://dx.doi.org/10.1016/B978-0-12-801740-1.00001-9

and diencephalic–telencephalic sulcus gives rise to five distinctive vesicles namely, the telencephalon, diencephalon, mesencephalon, metencephalon, and myelencephalon. Differing rates of cell division of each vesicle results in the transformation of a cylindrical neural tube into a more complex, folded structure. This in turn influences the size and shape of the cavities of each of the five divisions, which gives rise to their respective parts of the ventricular system. As the cerebral hemispheres continue to expand, so do the lateral ventricles, which are in close association. The lateral ventricles communicate with the single and narrow third ventricle, through the interventricular foramina of Monro. As the cells of mesencephalon continue to divide, the ventricular cavity is reduced in size giving rise to a narrowed cerebral aqueduct, which connects the third and fourth ventricle. With the closures of the rostral and caudal neuropores early in embryogenesis, the neural tube space gives rise to an enclosed ventricular system. By the third month of fetal development, foramina appear within the roof of the fourth ventricle that allows communication between the once closed ventricular system and the surrounding subarachnoid space. As the layer of connective tissue and ependymal cells that line the fourth ventricle, begin to break down, this gives rise to the formation of the three openings: a single, medial foramen of Magendie and two, lateral foramina of Luschka (Fig. 1.1).

Despite our increasing understanding of the development of the CNS, the function and purpose of the ventricular system is yet to be fully comprehended. Following the development of the embryonic forebrain, midbrain, and hindbrain ventricle formation, these ventricles expand at a much more rapid rate than brain tissue, thus making ventricle volume notably faster in growth.[1] Research on the molecular and cellular mechanism gives more insight into the brain ventricular system. Formation of the ventricles is dependent upon the neuroepithelium.[2] The surrounding neuroepithelium gives position and shape to the developing embryonic brain ventricular system. The neuroepithelium is arranged along the anteroposterior axis. With this pattern of placement, correct positioning of the ventricles is allowed and morphogenesis of the brain tissue is directed downstream. The arrangement of neuroepithelium occurs before and during neurulation.[2] During this period, embryonic brain tissue is subdivided into various gene expression domains. Patterning genes are responsible for the positioning of brain ventricles, including the characteristic as well as conserved constrictions and bends

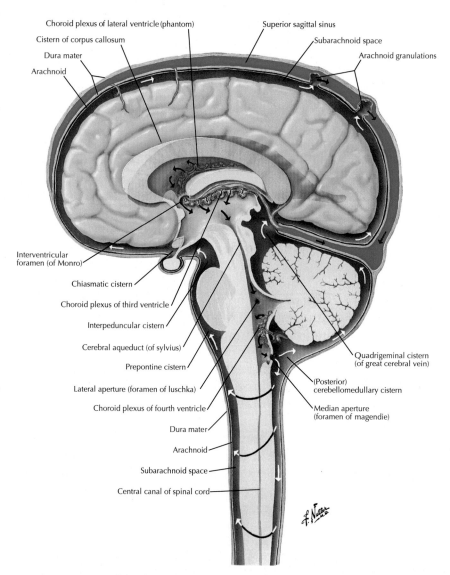

Choroid plexus of lateral ventricle (phantom)

Cistern of corpus callosum

Dura mater

Arachnoid

Superior sagittal sinus

Subarachnoid space

Arachnoid granulations

Interventricular foramen (of Monro)

Chiasmatic cistern

Choroid plexus of third ventricle

Interpeduncular cistern

Cerebral aqueduct (of sylvius)

Prepontine cistern

Lateral aperture (foramen of luschka)

Choroid plexus of fourth ventricle

Dura mater

Arachnoid

Subarachnoid space

Central canal of spinal cord

Quadrigeminal cistern (of great cerebral vein)

(Posterior) cerebellomedullary cistern

Median aperture (foramen of magendie)

FIGURE 1.1 CSF flow through the ventricular system. CSF produced from the choroid plexus flow from the lateral ventricles to the third ventricle through the interventricular foramina of Monro. From the third ventricle, CSF flows through the cerebral aqueduct and into the fourth ventricle. From here, CSF can continue further into the central canal of the spinal cord or into the subarachnoid space through the foramen of Magendie and foramina of Luschka. *Netter medical illustration used with permission of Elsevier Inc. Copyright 2016. All rights reserved. www.netterimages.com.*

within each region of the brain. The patterning genes may be responsible proximally in neuroepithelial morphogenesis, by having a control over the cytoskeletal machinery. On the other hand, they may have a distal role involving early tissue specification. Animal models have shed light on some of the mechanisms by this process takes place. An example of a distal role is when zebra fish embryos, lacking expression of fibroblast growth factor-8 (FGF8) were examined, an abnormal midbrain shape, along with improper shaping of the ventricles, and no presence of midbrain–hindbrain boundaries were observed.[2] In chick brains, Sonic Hedgehog (Shh) might have a proximal role in brain ventricle expansion. Shh is secreted from the notochord; if the notochord and brain are separated following initial patterning events, then ventricular expansion is prevented, which might show increased cell death and decreased cell proliferation. This may suggest that Shh plays a vital role in ventricle formation.[3]

The shape of the brain and ventricular cavities is determined by stereotypical and conserved morphogenesis, regulated cell proliferation, and cell death.[2] For brain morphogenesis and ventricle shaping to occur, the neuroepithelium must form intact and cohesively, along with the appropriate junctions. Tight junctions and apically localized adherens connect sheets of cells in embryonic neuroepithelium. Cells using these junctions create a division between the inside and outside of the neural tube. Again zebrafish mutants show a vital importance regarding tight junctions and their role during brain development. One example is the N-cadherin mutant where tight junctions do not form and neuroepithelial tissue fall apart. Sodium–potassium ATPase is a necessary ingredient for brain ventricle development. Three processes utilize this protein complex. First, the alpha subunit (Atp1a1) and the regulatory subunit Fyxd1 are required to form a cohesive neuroepithelium. Atp1a is a gene, which by itself only regulates neuroepithelial permeability. Fyxd1 however does not change neuroepithelial permeability, which may suggest a role in neuroepithelium formation. Second, RhoA further regulates formation and permeability. Third, CSF production is RhoA-independent, which requires Atp1a1 and not Fyxd1. Thus, retention and production of CSF are required for ventricle formation. The correct shape, functional cytoskeleton, and extracellular matrix are all needed in order to have proper formations.

The final stage of ventricular development uses a key ingredient – embryonic cerebrospinal fluid (eCSF) is secreted by the neuroepithelium; it is used to inflate the ventricles. Study of zebrafish shows how formation of eCSF depends on the sodium–potassium ATPase ion pump. Mutated zebrafish, which are lacking activity of sodium–potassium ATPase pump, fails to inflate the brain ventricles. By using the pump as an osmotic gradient, fluid can move into the ventricle lumen.[4] Others have made a suggestion that pressure from within the lumen coming from the eCSF is vital for brain development.[2] Although in adults the choroid plexus plays a major role with the development of CSF, not much is known about its role in brain ventricle development. The choroid plexus has not even begun to develop until after several weeks of inflation in humans.[1] Therefore, the only available source of eCSF that is known of must be the neuroepithelium. It is possible that neuroepithelium may be involved in eCSF production since it surrounds the ventricles at the time of embryonic development.[2]

FORMATION OF THE CHOROID PLEXUS

The formation of the choroid plexus involves the ependymal cells that line the luminal surface of the ventricles as well as the delicate connective tissue layer of the pia mater that lies beneath. Together, these two layers are referred to as the tela choroidea. As arteries in the immediate proximity begin to invaginate into the tela choroidea, this produces a narrow groove called the choroid fissure. Collectively, the newly formed out pouching of the tela choroidea and the underlying vessels into the lumen of the ventricle become the primitive choroid plexus. As this structure continues to enlarge and migrate outward, villi begin to form, and within them exists vascularized connective tissue in the form of capillaries. The endothelium of these capillaries contain many fenestrations that are essential for the exchange of molecules between the vasculature and the surrounding interstitial fluid. This coordinated process of choroid plexus development has been illustrated in a number of animal studies over several decades.[5–8] In particular, the continued differentiation and migration of choroidal epithelial cells have been investigated. Liddelow et al. used a *Monodelphis domestica* model to further characterize the growth pattern of the choroid plexus epithelium during development.[9] By injecting these animals with thymidine analogue 5-bromo-2-deoxyuridine (Brdu) at various stages of postnatal development, they were able to quantify immunopositive choroid epithelial cells throughout development as well as examine the location of these cells within the lateral ventricles. Furthermore, double labeling with plasma-protein-directed antibodies allowed researchers to identify cells that were functionally mature in terms of plasma exchange. Results suggested a 10-fold increase of epithelial cells between birth and adulthood. Furthermore, migration of these epithelial cells was similar to that of a conveyer belt, originating from the dorsal side of the plexus and moving toward the ventral surface. Finally, by counting the number of cells that were double stained with CrdU and endogenous plasma proteins, the researchers concluded that functional capabilities of protein transfer by these epithelial cells are acquired post-mitotically, but the birth of these cells occurs as early as the third postnatal day.[9] These results underscored the critical role in which early choroidal epithelial cells play in maintaining the correct CSF composition and concentrations to ensure stability, homeostasis, and normal CNS functioning.

Other studies have examined the signaling pathways and transcription factors that determine cell differentiation during early development. Johansson et al. examined the role of Otx2 transcription factor and its effects on choroid plexus development.[10] Choroid plexus development was halted through gene silencing of the Otx2 coding. Moreover, changes in early CSF composition, such as altered protein content, were also demonstrated in these mice. Mice that exhibited deletion of Otx2 displayed a lack of choroid plexus in all brain ventricles. Furthermore, it was shown that the transcription factor was necessary at later stages of development in order to properly maintain choroid plexus cells of the hindbrain.[10] Thus, conveys the importance of transcription factors like Otx2 in regulating choroid plexus development.

The formation of the choroid plexus in the lateral, third, and fourth ventricles are largely the same. Choroid fissures form in superior aspects of the third and fourth ventricles, in addition to the medial walls of the lateral ventricles. With growth, there will eventually be a continuation of the choroid plexus of lateral ventricles and third ventricle through the interventricular foramen. Zagorska-Swiezy et al. investigated the microvascular structure of the lateral ventricle choroid plexus of human fetal using scanning electron microscopy.[11] The study used male fetuses during the gestation period of 20 weeks that exhibited no developmental or maternal disorders. Results show primary villa present on the choroid plexus, while the development of more complex and lengthened true villa had not yet occurred. In addition, the anterior and posterior segments, the glomus and the villous fringe can be distinguished mainly because they display variations in vascular pattern. As a complete cast, the choroid plexus appeared as a thin, demilunar form with its concave being in close proximity with a thalamus-attached margin. It also exhibited a free margin area. The choroidal veins and arteries were mostly surrounded by capillary networks. Overall, the choroid plexus at the gestation period of 20 manifested the same similar patterns of mature choroid plexus except the absence of complex secondary villi.[11]

Cuboidal ependymal cells line the lumen of the entire ventricular system including the central canal of the spinal cord. These metabolically active cells are interconnected with their neighboring cells via zonulae adherens or desmosomes. The apical surface is characterized by cilia and microvilli, while astrocytic processes are in contact with the basal surface. Also within the ependymal layer exists specialized cells known as tanycytes. Although similar to ependymal cells, tanycytes also have basal processes that extend beneath the connective tissue layer and abut the underlying blood vessels. Moreover, tanycytes are connected to adjacent tanycytes and ependymal cells via tight junctions in addition to the desmosomes. Henson et al. used zebrafish to analyze the genetic elements for the development of the choroid plexus.[12] They generated enhanced green fluorescent protein in the epithelium of the myelencephalic and diencephalic choroid plexus by creating an enhancer trap line that would allow the analysis of development patterns. The choroid plexus included many occludin and claudin proteins that are involved in forming tight junctions with other adhesion proteins. In addition, rhodamine 123 was injected into the zebrafish enhancer trap line where it was observed to gather within the epithelium of the choroid plexus. This indicated the presence of transporter proteins. Additionally, choroid plexuses that exhibited abnormal formations were sequestered and some of the recessive mutants were shown to originate from chromosomes 4 and 21.[12] This allows the identification of genes, which are vital for the normal development of the choroid plexus and aid in understanding factors that contribute to disease.

Following early development, the final distribution of the choroid plexuses within the ventricular system has been extensively characterized. Within the lateral ventricles, choroid plexuses can be found distributed along the floor and medial wall. The atrium of the lateral ventricles, an area between the posterior and inferior horns, house particularly large clumps of choroid plexus called the glomus choroideum. Continuing further through the interventricular foramina, the choroid plexus extends into the third ventricle. Distribution of choroid plexuses within the third ventricle is limited to the tela choroidea located at the roof of the ventricle. Although in communication with the third and fourth ventricles, the cerebral aqueduct does not house any choroid plexus. At last, in the fourth ventricle the choroid plexus can be found along the tela choroidea of the roof and lateral aspects of the ventricle. The fourth ventricle contains the foramina, Luschka and Magendie, which are perhaps more important than the production of CSF, as they are the only exit points for the CSF to reach the subarachnoid space. Moreover, the caudal-most point of the ventricle narrows and continues as the central canal of the spinal void of the choroid plexus.

CEREBROSPINAL FLUID

The continuous production of CSF remains a vital component of CNS homeostasis. Production occurs at a rate of 10–20 mL/h with a total of 400–500 mL produced in a single day. The ventricular system is capable of housing anywhere between 120 mL and 150 mL, which indicates that in a given day, the total volume of CSF can be turned over three to four times. Choroid epithelial cells by means of molecular transport from surrounding extracellular fluid and connective tissue layers produce CSF. Much like the renal tubules, an array of asymmetrically positioned ion transporters allow fluid secretion into the ventricles (Fig. 1.2). Sodium enters from plasma ultrafiltrate or interstitial fluid across the basolateral surface via sodium–hydrogen ion exchangers down its concentration gradient. Subsequently sodium can be actively transported by sodium–potassium ATP pumps through the apical surface into the lumen of the ventricles. A study done by Amin et al. involved isolating the choroid plexus from young male rats and analyzing its epithelial sodium channels.[13] Some of the choroid plexus were treated with epithelial sodium channel blocker, benzamil, while others with ouabain, a sodium–potassium ATPase blocker. Results show that benzamil has a significant effect on sodium concentrations in the cell by inhibiting the epithelial sodium channels and stopping the influx of sodium, while sodium efflux still functions.

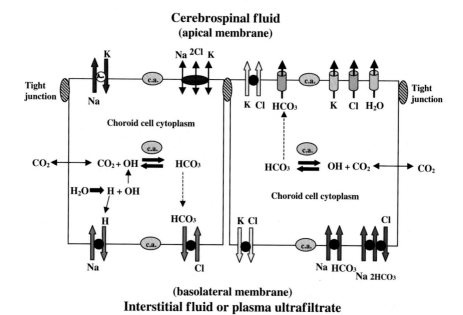

Cerebrospinal fluid
(apical membrane)

(basolateral membrane)
Interstitial fluid or plasma ultrafiltrate

FIGURE 1.2 **Ion channels and exchangers located along the apical and basolateral membrane of the choroidal epithelium.** The location of these key ion transporters and the cells overall polarity allows appropriate CSF secretion across the cell and into the lumen of the ventricles. *Reprinted with permission from Ref. [21].*

This brings a decrease in sodium concentration in choroid plexus cells. Ouabain, however, represses efflux and causes an increase in the retention of sodium in the choroid plexus. Sodium–potassium ATP pumps at the choroid plexus provides sodium emissions into the CSF. This indicates how sodium concentrations can be reduced by blockage.[13] Precise salt concentration in the CSF is critical in order to maintain proper functioning of neurons and the CNS. These results demonstrate how a complex and intricate transport mechanism is at play to ensure appropriate CSF compositions of sodium.

On the other hand, chloride is actively transported across the basolateral surface for bicarbonate and can diffuse across the apical surface, as the main anion in CSF. Bicarbonate formation occurs intracellularly during hydration of carbon dioxide catalyzed by carbonic anhydrase. In addition, bicarbonate can also enter the choroidal epithelial cell through sodium-coupled bicarbonate transporters. Intracellular bicarbonate can then diffuse down its concentration gradient by either an anion channel embedded in the basolateral membrane or by a sodium-coupled bicarbonate co-transporter. Amino acids also exist in the CSF in order to maintain homeostasis and allow normal neurological functioning. Kitazawa et al. used transgenic rats containing a simian virus 40 T-large anti-gen gene to further characterize amino acid transport.[15] This model was used as a source of immortalized cell line. The large T-antigen gene becomes expressed in all animal tissues and the cell cultures can be readily preserved. Choroid plexus epithelial cells were se-questered from simian virus 40 T-large antigen gene rats to depict the transport actions and epithelial markers of the choroid plexus. Transthyretin, specific thyroxine transport protein, was revealed in the choroid plexus epithelial cell line with sodium–potassium ATPase located at the apical side along brush borders by the CSF of the choroid plexus. In addition, choroid plexus epithelial cell lines were shown to be polarized since there was a significant amount of L-proline uptake at the apical side compared to the basal end. This denotes how active transport mechanisms direct the efflux of L-proline by the CSF side. Furthermore, small neutral amino acids hindered the uptake of L-proline by choroid plexus epithelial cells.[15] These specific transport mechanisms are vital to sustain a constant concentration of amino acids in the CSF.

The CSF also consists of many organic substances, such as pep-tides and drugs, which get influxed through the blood–brain bar-rier by specific membrane transporters. Studies have determined that two organic anion transporting polypeptides, Oatp1 and Oatp2, isolated from rat brain play a key role in the movement of many organic amphipathic substrates and also in drug transport.[16] Oatp1, localized at the apical surface, and Oatp2, in the basolateral region, convey their complementary role in regulating CSF com-position.[17] The polar localization of these two organic transporting polypeptides in the choroid plexus epithelium also allow drugs to target brain tissue and provide therapeutic effects. With the move-ment of these osmotically active molecules across the blood–CSF barrier, water can then follow through aquaporin channels located on the apical surface of the cell. Oshio et al. used targeted gene distribution to generate aquaporin-1 null mice in order to compare differences of water permeability between wild type mice that ex-hibit aquaporin-1 versus genetically modified mice that have an absence of aquaporin-1.[18] Aquaporin-1 acts as a water pore to aid in the movement of water across the epithelium due to osmotic gradi-ent. Results indicate that there is a fivefold reduction in the osmotic to diffusional water ratio penetrability of choroidal epithelium in mice lacking aquaporin-1. This decrease in permeability implies that fewer total amounts of water will be osmotically moved across the ventricle side of the choroid epithelium by transmembrane transport systems. Mice with a deletion of aquaporin-1 had lower amount of CSF production as a consequence. This denotes that aquaporin-1 transports a significant amount of water and plays an immense role in CSF production.[18,19] As a result, the composition of CSF is 99% water, which further underscores the importance of functional aquaporin channels allowing adequate CSF production.

VASCULAR SUPPLY OF THE CHOROID PLEXUS

The choroidal arteries are the main vascular support systems for the choroid plexus. There are two main choroidal arteries: the anterior choroidal arteries (AChA) and the posterior choroidal arteries (PChA). The AChA is the vessel that supplies vascular support to the choroid plexus in the lateral and third ventricles. It also provides support to the posterior limb of the internal capsule and the optic radiation. This artery originates as a branch off of the internal carotid artery (ICA), although there can exist variations to the branch off point. The intraoperative study of Akar et al. examined variations in AChA anatomy and they found that 70% arose from the ICA distal to the posterior communicating artery (PCoA), 20% from the origin of the PCoA, and 10% from the ICA bifurcation.[20] The AChA, like all blood vessels associated with the brain, are at a risk of aneurysms. Fortunately, AChA aneurysms only account for 4% of intracranial aneurysms, but should complications such as ruptures occur, this can lead to devastating injuries. Occlusions of the AChA classically appear symptomatic as contralateral hemiplegia, hemianesthesia, and hemianopsia.

The posterior choroidal arteries can be further divided into medial and lateral divisions. The medial posterior choroidal arteries branch from the posterior cerebral artery (PCA) or its smaller branches, such as the calcarine and parieto–occipital branches, to supply portions of the midbrain, thalamus, pineal region, and the choroid plexus of the third ventricle before anastomosing with the lateral posterior choroidal artery. The lateral posterior choroidal artery also branches from the PCA distal to the medial branch and supplies among other structures the choroid plexus of the lateral ventricles. The vascular supply to the choroid plexus of the fourth ventricle can be varied as Sharifi et al. found in their cadaveric model that the anterior inferior cerebellar artery (AICA), posterior inferior cerebellar artery (PICA), and superior cerebellar artery (SCA), all have contributing roles.[21] Should the blood flow in any of these vessels be impeded, significant morbidity and mortality may occur as a result of ischemic injury or increased intracranial pressure, necessitating prompt medical or surgical intervention.

References

1. Bayer SA, Altman J. *The Human Brain During the Early First Trimester. Atlas of Human Central Nervous System Development.* Boca Raton, FL: CRC; Taylor & Francis distributor; 2008:522 p.
2. Lowery LA, Sive H. Initial formation of zebrafish brain ventricles occurs independently of circulation and requires the nagie oko and snakehead/atp1a1a.1 gene products. *Development.* 2005;132(9):2057–2067.
3. Britto J, Tannahill D, Keynes R. A critical role for sonic hedgehog signaling in the early expansion of the developing brain. *Nat Neurosci.* 2002;5(2):103–110.
4. Brown PD, Davies SL, Speake T, et al. Molecular mechanisms of cerebrospinal fluid production. *Neuroscience.* 2004;129(4):957–970.
5. Keep RF, Jones HC. A morphometric study on the development of the lateral ventricle choroid plexus, choroid plexus capillaries and ventricular ependyma in the rat. *Brain Res Dev Brain Res.* 1990;56(1):47–53.
6. Stastny F, Rychter Z. Quantitative development of choroid plexuses in chick embryo cerebral ventricles. *Acta Neurol Scand.* 1976;53(4):251–259.
7. Sturrock RR. A morphological study of the development of the mouse choroid plexus. *J Anat.* 1979;129(Pt. 4):777–793.
8. Tennyson VM, Pappas GD. Fine structure of the developing telencephalic and myelencephalic choroid plexus in the rabbit. *J Comp Neurol.* 1964;123: 379–411.
9. Liddelow SA, Dziegielewska KM, Vandeberg JL, et al. Development of the lateral ventricular choroid plexus in a marsupial, *Monodelphis domestica. Cerebrospinal Fluid Res.* 2010;7:16.
10. Johansson PA, Irmler M, Acampora D, et al. The transcription factor Otx2 regulates choroid plexus development and function. *Development.* 2013; 140(5):1055–1066.
11. Zagorska-Swiezy K, Litwin JA, Gorczyca J, et al. The microvascular architecture of the choroid plexus in fetal human brain lateral ventricle: a scanning electron microscopy study of corrosion casts. *J Anat.* 2008;213(3):259–265.
12. Henson HE, Parupalli C, Ju B, et al. Functional and genetic analysis of choroid plexus development in zebrafish. *Front Neurosci.* 2014;8:364.
13. Amin MS, Reza E, Wang H, et al. Sodium transport in the choroid plexus and salt-sensitive hypertension. *Hypertension.* 2009;54(4):860–867.

14. Kitazawa T, Hosoya K, Watanabe M, et al. Characterization of the amino acid transport of new immortalized choroid plexus epithelial cell lines: a novel *in vitro* system for investigating transport functions at the blood-cerebrospinal fluid barrier. *Pharm Res.* 2001;18(1):16–22.

15. Noe B, Hagenbuch B, Stieger B, et al. Isolation of a multispecific organic anion and cardiac glycoside transporter from rat brain. *Proc Natl Acad Sci USA.* 1997;94(19):10346–10350.

16. Gao B, Stieger B, Noé JM B, et al. Localization of the organic anion transporting polypeptide 2 (Oatp2) in capillary endothelium and choroid plexus epithelium of rat brain. *J Histochem Cytochem.* 1999;47(10):1255–1264.

17. Oshio K, Song Y, Verkman AS, et al. Aquaporin-1 deletion reduces osmotic water permeability and cerebrospinal fluid production. *Acta Neurochir.* 2003;86(suppl.):525–528.

18. Oshio K, Watanabe H, Song Y, et al. Reduced cerebrospinal fluid production and intracranial pressure in mice lacking choroid plexus water channel Aquaporin-1. *FASEB J.* 2005;19(1):76–78.

19. Akar A, Sengul G, Aydin IH. The variations of the anterior choroidal artery: an intraoperative study. *Turk Neurosurg.* 2009;19(4):349–352.

20. Sharifi M, Ciołkowski M, Krajewski P, et al. The choroid plexus of the fourth ventricle and its arteries. *Folia Morphol Warsz.* 2005;64(3):194–198.

21. Johanson CE, Duncan III JA, Klinge PM, et al. Multiplicity of cerebrospinal fluid functions: new challenges in health and disease. *Cerebrospinal Fluid Res.* 2008;5:10.

Development of Brain Ventricles and Choroid Plexus

Ellen M. Carpenter

Department of Psychiatry and Biobehavioral Sciences and
Interdepartmental Program in Neuroscience, David Geffen School
of Medicine, University of California, Los Angeles, CA, USA

INTRODUCTION

Between our ears, we hold a complicated, integrated, and mysterious piece of tissue. Regulating our movement, language, emotions, and perception, the human brain is perhaps the most complex structure in the known universe. Yet, like all elements of the body, the brain and spinal cord, which together constitute the central nervous system (CNS), come from humble origins. At the start of human development, a single fertilized cell divides to produce a small clump of amorphous, interchangeable cells. Cell movements organize this piece of tissue into a two-layered mass, with a top and a bottom, then through purposeful, directed migration, cells from the

The Choroid Plexus and Cerebrospinal Fluid. http://dx.doi.org/10.1016/B978-0-12-801740-1.00002-0

top of the structure migrate between the layers, producing a third layer. These three embryonic germ layers, the ectoderm, mesoderm, and endoderm, are the raw materials needed to build an entire human. The nervous system, along with the skin, is a derivative of the ectoderm, while the mesoderm gives rise to the skeleton and musculature, and the endoderm produces the majority of the gastrointestinal tract. This chapter will examine the early steps in the specification of neural tissue from undifferentiated ectoderm, with a specific focus on the development of the ventricular system and choroid plexus.

NEURAL PLATE AND NEURAL TUBE FORMATION

The vertebrate nervous system arises along the dorsal aspect of developing embryos as a derivative of the ectoderm. Nearly 100 years ago, the pioneering work of Spemann and Mangold[1] demonstrated that initial instructions for the production of the nervous system stemmed from signals arising outside the nervous system, in the dorsal mesoderm. By transplanting small bits of tissue isolated from amphibian embryos in the region, where cells involutes to form the mesoderm, to distant locations near the belly of the host animals, Spemann and Mangold were able to demonstrate the production of a second nervous system and surrounding axial structures. When they examined the region of the transplant, they were able to see that the transplanted cells had formed dorsal mesoderm, and that the surrounding host tissue had been induced to form a new nervous system. Spemann and Mangold reasoned that the transplanted tissue, which they named the "organizer", had instructed the surrounding host tissue to form nervous tissue. This implied the presence of one or more molecular signals emanating from the organizer that could provide necessary instructions to build a nervous system. Although the term "organizer" is specifically applied to amphibians, homologous structures have since been identified in other organisms, including the shield in zebrafish, Hensen's node in chick embryos, and the node in mammalian embryos such as the mouse, rat, and human.

The initial observations of Spemann and Mangold[1] led to a number of studies focused on identifying the molecule(s) that provide the necessary signals to establish the nervous system. Early experiments demonstrated that a large number of physiological and nonphysiological factors could stimulate the formation of nervous tissue.[2] Heat treatment could destroy many physiological factors, suggesting that such factors were likely to be proteins. Initial assumptions implied that the organizing signal provided active instruction for the induction of the nervous system, but later studies suggested that neural development might in fact function by disrupting an active signaling mechanism. This was demonstrated by examining pieces of naïve embryonic ectoderms that were grown in tissue culture. When the structure of the explants was maintained and contacts between adjacent cells were left intact, the explants formed the epidermis. However, when the explants were dissociated and single cells were cultured, the cells developed a neural character and expressed neural-specific molecules. This suggested that the default pathway of development for these ectodermic cells was to become neural, while there was an active mechanism required for the cells to develop into an epidermis.

With the realization that molecular signals were required for the development of epidermal fates, the hunt was on for the signaling pathways that directed this phase of development. This rapidly led to an examination of signaling by TGF-β family members. Some bone morphogenetic proteins (BMPs), a large subgroup of TGF-β family members, were known to be expressed in the developing ectoderm, and several studies had demonstrated that known antagonists of BMP signaling or dominant-negative BMP receptors were capable of inducing isolated ectodermic cells to adopt epidermal fates.[3,4] A variety of molecules, including noggin, chordin, follistatin, and cerberus, which were known to bind to and block BMP signaling, were subsequently shown to be able to induce the formation of neural tissue when applied to naïve ectoderm.[2]

With these results, a model has been developed whereby BMP antagonists, including not only noggin, chordin, and follistatin, but others as well, are secreted from mesoderm tissue, and act to inhibit BMP signaling between adjacent cells in the ectoderm, diverting the cells from an epidermal to a neural fate.[4] Mesoderm cells involutes through the organizer, shield, or node aggregate to form a flexible, rod-like structure called the notochord that elongates along the anteroposterior axis of the developing embryo. Notochord cells continue to express the same inducing molecules, thus inducing development of neural competency in a strip of dorsal tissue overlying the notochord. This strip of tissue is the neural plate, the first anatomically recognizable precursor of the nervous system. Signaling in at least two dimensions is patterned by the neural plate. Vertical signals arising from the underlying notochord establish the position of the plate, while horizontal signals from the adjacent nonneural ectoderm establish the margins of the plate, and contribute to the development of dorsal cell fates, which will develop further during neurulation. Horizontal signals also contribute to the differentiation of the neural crest, a migratory population of cells arising from the early neural tube that populates the peripheral sensory and autonomic ganglia. Interestingly, some of the planar horizontal signals are provided by BMPs, suggesting that these molecules are inhibitory and instructive at different points in the development of the nervous system.

In response to inducing signals, ectodermic cells overlying the notochord lengthen along their apico-basal axis, becoming morphologically distinct from the surrounding cuboidal-epithelial cells and begin to express molecular markers characteristic of neural rather than ectodermic cells. The shape change allows morphological distinction of the neural plate from the surrounding epidermal cells, and the expression of neural-specific molecules demonstrates the restriction of this part of the ectoderm to a neural fate. This is the first step of the four steps of neurulation, which ultimately leads to the formation of the neural tube, the obligate precursor for the vertebrate nervous system. In the second step, the neural plate elongates rostrocaudally along the dorsal midline and narrows in the mediolateral axis through the process of convergent extension, with the exception of cells in the midline of the neural plate, which are anchored to the underlying notochord.[5] These midline cells become constricted at their apical edge, producing a wedge- or bottle-like shape, which starts the physical changes that lead to the formation of a rounded tube from a flat plate of cells. The third step in neurulation is bending, which involves the formation of dorsolateral hinge points in addition to the midline hinge point, raising of the neural folds at the lateral edges of the neural plate, folding of the neural plate into a tube, followed by the final step, fusion of the neural folds to form an enclosed neural tube. These movements restructure the neural plate from a flat sheet of elongated cells into a hollow tube comprised of a cell wall surrounding a central lumen. In humans and mice, neural tube fusion begins at the level of the hindbrain with the dorsal apposition of the neural folds and continues rostrally as well as caudally along the length of the embryo. During the process of neural tube fusion, the lumen of the neural tube remains in contact with the surrounding environment via the anterior and posterior neuropores. Once these structures close, the lumen is isolated from the rest of the embryo; this marks the beginning of the closed ventricular system. A failure in rostral neural tube fusion, either through incomplete fusion of the neural folds or through a failure in the closure of the anterior neuropore produces anencephaly, while a failure in caudal neural tube fusion or posterior neuropore closure produces *spina bifida*.

The molecular processes driving the cell shape changes involve active rearrangement of the cytoskeleton. Apicobasal extension involves microtubule polymerization and assembly parallel to the axis of the elongating cell; the PDZ domain-containing protein Shroom3, is a critical factor in this organization. Loss of Shroom3 results in the failure to assemble thick, apicobasally oriented microtubules.[6] The formation of wedge-shaped cells stems from thickening of actin filaments and contraction of nonmuscle myosin II localized at the apical edges of the neural plate cells; this process is also modulated by Shroom3, as loss of the protein reduces apical accumulation of F-actin. Structurally, Shroom3 is localized to adherens junctions that form between neural plate cells. The family of small Rho GTPases, including Rho, Rac, and Cdc42, are key regulators of cytoskeletal organization and inhibitors of Rho-associated kinases (ROCKs). This causes a reduction in phosphorylated myosin light chain and a failure in neural tube closure; interestingly, ROCKs physically interact with Shroom3 and this interaction is required for phosphorylation of myosin light chain. During embryogenesis, Rho activity is positively regulated by Wnt signaling, through the noncanonical planar cell polarity (Wnt/PCP) pathway. Cell adhesion molecules also play a role in cellular morphogenesis in neural tube closure. N-cadherin, a calcium-dependent cell adhesion protein is evident in the neural plate, with its expression concentrated apically. N-cadherin is required for the maintenance of cell surface tension and contractility, by linking F-actin to the cell surface. Disruption in the expression of this protein in Xenopus causes failure in neural tube closure. Disruption of N-cadherin expression in mice produces few neural tube defects; however, if Shroom3 and N-cadherin are disrupted, failures in neural tube closure are occasionally seen.

With the completion of neural tube fusion, typically by 28 days of gestation in human embryos, the precursor of the CNS is formed. At this stage, the walls of the neural tube are one cell layer thick and a bulk of the cells in the wall of neural tube are neural progenitor cells that are primed to proliferate and produce all the cells of the nervous system, namely neurons and glia. However, there are some specialized cells along the ventral and dorsal edges of the neural tube that are not neural progenitor cells. Ventrally, these cells make up the floor plate, and dorsally they are the roof plate. The floor plate and the roof plate serve as signaling centers that provide patterning information for the development of different cell types along the dorsoventral axis of the neural tube. Ventrally, the floor plate secretes sonic hedgehog (Shh), which regulates the development of motor neurons and ventral interneurons in a concentration-dependent fashion. Dorsally, the roof plate secretes several proteins, including BMPs that regulate the development of more dorsal cell types such as commissural neurons and dorsal interneurons. The floor plate and the roof plate are morphologically distinct, and can be easily identified in sections through the developing spinal cord.

When it first forms, the neural tube is essentially uniform along its anteroposterior axis. However, there is a rapid process of expansion and morphogenetic movement rostrally that produces three primary brain vesicles, the prosencephalon, the mesencephalon, and the rhombencephalon (known colloquially as forebrain, midbrain, and hindbrain). The appearance of these vesicles is heralded by an expansion of the central lumen and constriction of the walls of the neural tube at the boundaries between the different vesicles, along with bending of the neural tube. The first bend, the cephalic flexure, which bends the rostral part of the neural tube down and forward, occurs between the prosencephalon and mesencephalon, and the prosencephalon balloons out rapidly rostral to the flexure. More caudally, the cervical flexure, which also bends the neural tube forward, occurs at the level of the juncture between the rhombencephalon and the spinal cord. The three primary brain vesicles are subsequently divided into five secondary brain vesicles, with the prosencephalon splitting into telecephalon and diencephalon, the mesencephalon remains undivided, and rhombencephalon segments into metencephalon and myelencephalon. In concert with the formation of the secondary brain vesicles, a third flexure, the pontine flexure, also bends the neural tube; this flexure occurs in the opposite direction to cephalic and cervical flexures, serving to double the neural tube back on itself. The flexures are necessary to allow the expanding neural tube to fold to fit inside the surrounding skull. The five secondary brain vesicles are the forerunners for the structures of the adult vertebrate brain, with the telencephalon giving rise to the cerebral hemispheres, the diencephalon giving rise to the thalamus, hypothalamus, and optic cups, the mesencephalon giving rise to the midbrain, the metencephalon giving rise to the pons and cerebellum, and the myelencephalon giving rise to the medulla oblongata.

VENTRICULAR SYSTEM DEVELOPMENT

Once the anterior and posterior neuropores have closed, lumen of the neural tube is sealed off from the surrounding environment. This lumen is the precursor of the ventricular system, which constitutes a closed circulatory system inside the brain. In the adult mammalian brain, there are four main ventricles – the paired lateral ventricles, the third ventricle, and the fourth ventricle. The lateral ventricles, which begin as a single central ventricle, form rostrally, inside the telencephalon. As the progenitor cells in the walls of the neural tube proliferate, the telencephalon expands laterally over the surface of the more caudal brain vesicles. The lumen inside the telencephalon follows this lateral expansion to form the lateral ventricles, which are connected to the third ventricle via the interventricular foramena. Ventromedial expansion of the medial and lateral ganglionic eminences invades the lateral ventricles, generating the familiar C-shape. The third ventricle is largely formed in the diencephalon, with a small rostral extension into the midline of the telencephalon. The neural tube lumen is constricted through the mesencephalon, forming the cerebral aqueduct, which then expands into metencephalon and myelencephalon to form the rhombus-shaped fourth ventricle. The fourth ventricle is continuous with the central canal of the spinal cord. Thus, the CNS, which began life as a tube, retains its tubular structure through the presence of the internal ventricular system. The ventricular system contains cerebrospinal fluid (CSF), which is secreted into the system by the choroid plexi found within each ventricle; the ventricular system thus forms an enclosed circulatory space for CSF within the brain.

The ventricular system is connected to the subarachnoid space through several foramena below the cerebellum. This allows circulation of the CSF from within the ventricles into cisterna magna, returning to the peripheral circulation by resorption through the arachnoid granulations in the venous sinuses of the brain. Together, this constitutes an independent circulatory system for the brain and spinal cord.

During brain development, cell proliferation occurs adjacent to the ventricles in a region known as the ventricular zone. Postmitotic neurons exit the ventricular zone and migrate away from the ventricles to populate regions of the brain. Initially, as all cells are proliferating, the ventricular zone spans the entire width of the neural tube. As proliferation decreases and the brain matures, the ventricular zone becomes thinner, and in the adult brain, proliferation is restricted to a few neurogenic regions, specifically the subventricular zone of the lateral ventricles, where neurons are generated that migrate via the rostral migratory stream into the olfactory bulb, and the subgranular zone of the dentate gyrus in the hippocampus. As the ventricular zone disappears, a layer of ependymal cells is formed around the ventricular system. Ependymal cells are postmitotic cells that are derived from radial glia. These cells are cuboidal and multiciliated; the cilia play an essential role in the propulsion of CSF through the ventricular system. Malfunctions of the cilia can lead to hydrocephalus. Ependymal cells serve as barriers modulating the flow of material from the brain parenchyma into or out of the CSF. In mice, a majority of ependymal cells are generated between embryonic day (E) 14 and 16, approximately two-thirds of the way through gestation. Immature ependymal cells are not ciliated; cilia appear shortly after birth.[7] Ependymal cells lining the ventricles are linked by desmosomes, which are specialized intercellular adhesion sites enabling the cells to form a nearly continuous epithelial sheet lining the ventricles. Desmosomes are loose junctions, allowing transfer of CSF (or components of the CSF) into the brain. During the embryonic period, CSF serves as a reservoir for growth factors and hormones needed for normal brain development and the ependymal cells allow transfer of these molecules into the developing brain. Composition of the CSF changes in adulthood, and some growth factors or other developmentally essential molecules are found in adult CSF. Tanycytes are rare cells that are found in the ependymal lining of the floor of the third ventricle. Tanycytes have long processes and large-end feet that connect to brain capillaries and neurons distant from the ventricle. Tanycytes facilitate the transfer of hormones and other substances from the CSF to and from neurons and capillaries.

CHOROID PLEXUS DEVELOPMENT

The mammalian choroid plexus is a highly vascularized tissue that develops within each of the brain ventricles. The choroid plexus epithelium is in direct contact with the ventricles, allowing ready access to the fluid circulation in the ventricles. About 100 years ago, the choroid plexus was discovered to be the source of CSF.[8] Prior to this discovery, the CSF was not considered an integral part of the ventricular and circulatory system of the brain, but possibly a postmortem artifact. In humans, the choroid-plexus-epithelial cells secrete approximately 400–600 mL of CSF each day, enough to turn over the CSF three to four times daily. Like all epithelia, the choroid plexus epithelium is composed of polarized cells, with basally located nuclei, apically enriched mitochondria, extensive luminal microvilli, and juxtaluminal tight junctions. In addition, the integral membrane protein aquaporin, which forms water-conducting pores, is preferentially localized to the microvilli on the luminal surface.

The choroid plexus is present in each of the four brain ventricles. In humans, the choroid plexus first appears in the fourth (hindbrain) ventricle, then the lateral ventricles, and finally in the third ventricle. There is evidence in mammalian species that the third ventricle choroid plexus is continuous with the choroid plexus in the lateral ventricles and may thus be derived from the same embryonic origin. The hindbrain choroid plexus is spatially segregated from the lateral ventricle and third ventricle choroid plexi and likely arises from a separate embryonic territory. The choroid plexus forms adjacent to the dorsal midline in the ventricles. There is evidence to suggest that choroid plexus progenitors are segregated very early in development, as explant culture studies have demonstrated that choroid plexus cells can only differentiate from distinct regions of the neural ectoderm at early embryonic stages (E8.5) in mice.[9] Anatomically, the first evidence of choroid plexus development is the differentiation of columnar neuroepithelial cells into a more cuboidal epithelial morphology, a reversal of the early developmental processes that initially produced elongated neuroepithelial cells from the cuboidal ectodermic cells. This is evident as early as E12.5 in mice, or by 6 weeks of gestation in humans.[9–11] Morphological differentiation is accompanied by the expression of the plasma thyroid transport protein transthyretin, which serves as a definitive molecular marker for committed choroid plexus epithelial cells. Stromal cells are recruited from the mesenchyme adjacent to the differentiating epithelium and as development progresses, the region becomes highly vascularized.

The choroid plexus as a whole is composed of the choroid plexus epithelial layer adjacent to the ventricle, a basal lamina, to which the epithelial cells are anchored, and an inner core of mesenchymally derived stromal cells surrounding a dense vascular network. This arrangement serves to separate the vasculature from ventricular system, but allows regulated transfer of materials from one compartment to the other. The choroid plexus may be considered a part of the blood–brain barrier, along with the lining of the vasculature of the CNS, as both serve to segregate the contents of the vasculature from the brain parenchyma and the CSF. However, there are significant differences between these two systems. The vascular blood–brain barrier is composed of endothelial cells lining the vasculature. These cells are derived from the endoderm, in contrast to the epithelial tissue in the choroid plexus, which come from ectodermally derived neuroepithelial tissue. Vascular endothelial cells line all blood vessels in the body and uniformly proliferate. This allows local repair of the vasculature, but can also be hijacked during tumor angiogenesis to line tumor-infiltrating capillaries. Pericytes, which are related to smooth muscle cells and are part of the connective tissue family, are scattered outside the vascular endothelial cells; these cells may mediate instructions relating to the development of the vasculature during choroid plexus development. Vascular endothelial cells are linked by tight junctions that are also seen within the choroid plexus epithelium, in contrast to the desmosomes that connect the ependymal cells lining the remainder of the vasculature. Tight junctions regulate the transfer of material across cells; this serves to separate the vasculature from the brain parenchyma (in the case of the vascular endothelial cells) or from the CSF (in the case of the choroid plexus epithelium).

Genetic lineage analysis suggests that choroid plexus epithelial cells originate from the roof plate of the neural tube and choroid plexus epithelial cell differentiation is evident even before neurogenesis begins or ependymal differentiation occurs; choroid plexus epithelial cell progenitors may be specified as early as E8.5 in the mouse. BMP signaling, in particular through BMP4, is required for choroid plexus formation. Deletion of BMP receptor expression in the dorsal telencephalon eliminates early differentiation of the choroid plexus epithelium. Observation throughout later embryonic stages demonstrates a complete absence of choroid plexus.

Choroid plexus epithelial cell development can be partially restored in roof plate-ablated CNS explants by treatment with BMP4.[12] Interestingly, loss of BMP signaling appears not to affect the production of choroid plexus progenitors, but it does impede their differentiation, as judged by a lack of transthyretin expression. Recent studies have also shown that BMP4 is sufficient to induce choroid plexus epithelial cell formation from embryonic stem cell-derived neuroepithelial tissue. In this study, pluripotent embryonic stem cells were first differentiated to neuroepithelial cells, then treatment with BMP4 was sufficient to induce a small percentage of these cells to develop epithelial polarity and to express genes, including Ttr, Msx1, Aqp1, Cldn1, and Lmx1a, that are characteristic of the developing choroid plexus epithelium.[13]

In the hindbrain, Shh is an instructive signal for choroid plexus development, particularly for regulating a match between epithelial and vascular components. Shh signaling is unique to hindbrain choroid plexus development, as Shh expression is not observed in the developing choroid plexi in lateral or third ventricles. Shh appears to regulate the proliferation of hindbrain choroid plexus epithelial cell progenitors and coordinated development of the vasculature. Shh expression is required for biogenesis of choroid plexus, with loss of signal leading to a reduction in epithelial and mesenchymal mass and in vascularity.[14,15] In the hindbrain, Shh is expressed by choroid plexus epithelial cells, while Shh receptor Ptch1 and Shh effector Gli1, are expressed in pericytes in the adjacent mesenchyme. A recent study has shown that Shh has no direct effect on vascular endothelial cells, but instead regulates vascular growth through communication with the pericytes that in turn instructs vascular endothelial cells.[15] In addition to signaling from the epithelium to the mesenchyme, Shh may be required to maintain a population of choroid plexus progenitors in the epithelium. There is a small group of Shh-expressing cells in a restricted domain near the dorsal midline that also express the Shh effector, Gli1. In the absence of Shh expression, cell proliferation decreased in this zone, leading to a reduced population of choroid plexus epithelial cells.[14] Thus, Shh produced in the choroid plexus epithelium not only signals to the mesenchyme, but also to its own progenitor domain. Interestingly, Shh is widely recognized as a signal for ventral neural tube development that is produced by the notochord and floor plate. An instructive role for Shh in choroid plexus development indicates that the molecule may have dorsal functions as well.

The functions of choroid plexus, particularly its permeability to proteins and low molecular weight compounds, changes between embryonic development and adulthood. This is evident if one examines the composition of CSF at different ages. In embryos, the CSF contains morphogens, mitogens, and trophic factors that assist in patterning early brain development. Embryonic CSF also contains a higher concentration of plasma proteins than is seen in adult CSF. Embryonic CSF has the capacity to support and promote growth of neural-stem cells and cortical explants, while adult CSF does not have these abilities.[16,17] These changes in CSF composition support a shift in the permeability of choroid-plexus-epithelial cells between embryonic and adult stages, allowing entry of molecules needed for brain development at the appropriate age and excluding these molecules once the process of development is complete.

In summary, the ventricular system and associated choroid plexus develop as an integral part of the CNS. The ventricular system maintains a tubular organization of the neural tube and encloses a central circulatory space for the CSF, which is secreted by the choroid plexus. The choroid plexus, which is composed of epithelial and stromal compartments, provides a regulatory barrier to the entry of blood and plasma components into the CSF; transfer of these materials is modulated between embryonic and adult stages, by changes in the permeability of choroid-plexus-epithelial cells.

References

1. Spemann H, Mangold H. Induction of embryonic primordia by implantation of organizers from a different species. *Roux's Arch Entw Mech.* 1924;100:599–638:(reprinted and translated in Int J Dev Biol 45: 13–31, 2001).
2. De Robertis EM. Spemann's organizer and the self-regulation of embryonic fields. *Mech Dev.* 2009;126:925–941.
3. Rogers CD, Moody SA, Casey ES. Neural induction and factors that stabilize a neural fate. *Birth Defects Res C.* 2009;87:249–262.
4. Munoz-Sanjuan I, Brivanlou AH. Neural induction, the default model and embryonic stem cells. *Nat Rev Neurosci.* 2002;3:271–280.
5. Smith JL, Schoenwolf GC. Neurulation: coming to closure. *Trends Neurosci.* 1997;20:510–517.
6. Suzuki M, Morita H, Ueno N. Molecular mechanisms of cell shape changes that contribute to vertebrate neural tube closure. *Dev Growth Differ.* 2012;54: 266–276.
7. Spassky N, Merkle FT, Flames N, Tramontin AD, Garcia-Verdugo JM, Alvarez-Buylla A. Adult ependymal cells are postmitotic and are derived from radial glial cells during embryogenesis. *J Neurosci.* 2005;25:10–18.

8. Cushing H. Studies on the cerebro-spinal fluid. *J Med Res*. 1914;31:1–19.
9. Thomas T, Dziadek M. Capacity to form choroid plexus-like cells *in vitro* is restricted to specific regions of the mouse neural ectoderm. *Development*. 1993;117:253–262.
10. Dziegielewska KM, Ek J, Habgood MD, Saunders NR. Development of the choroid plexus. *Microsc Res Tech*. 2001;52:5–20.
11. Hébert JM, Mishina Y, McConnell SK. BMP signaling is required locally to pattern the dorsal telencephalic midline. *Neuron*. 2002;35:1029–1041.
12. Cheng X, Hsu CM, Currie DS, Hu JS, Barkovich AJ, Monuki ES. Central roles of the roof plate in telencephalic development and holoprosencephaly. *J Neurosci*. 2006;26(29):7640–7649.
13. Watanabe M, Kang Y-J, Davies LM, et al. BMP4 sufficiency to induce choroid plexus epithelial fate from embryonic stem cell-derived neuroepithelial progenitors. *J Neurosci*. 2012;32:15934–15945.
14. Huang X, Ketova T, Fleming JT, et al. Sonic hedgehog signaling regulates a novel epithelial progenitor domain of the hindbrain choroid plexus. *Development*. 2009;136:2535–2543.
15. Neilson CM, Dymecki SM. Sonic hedgehog is required for vascular outgrowth in the hindbrain choroid plexus. *Dev Biol*. 2010;340:430–437.
16. Lehtinen MK, Zappaterra MW, Chen X, et al. The cerebrospinal fluid provides a proliferative niche for neural progenitor cells. *Neuron*. 2011;69:893–905.
17. Lehtinen MK, Bjornsson CS, Dymecki SM, Gilbertson RJ, Holtzman DM, Monuki ES. The choroid plexus and cerebrospinal fluid: emerging roles in development, disease and therapy. *J Neurosci*. 2013;33:17553–17559.

Choroid Plexus: Structure and Function

Florence M. Hofman[*,†],
Thomas C. Chen[†]

*Department of Pathology, Keck School of Medicine, University of Southern California, Los Angeles, CA, USA; [†]Department of Neurological Surgery, Keck School of Medicine, University of Southern California, Los Angeles, CA, USA

BASIC ANATOMY

The choroid plexus (CP) is the organ localized to the lining of the brain ventricles, responsible for providing a selective gateway for immune cells and specific agents to enter the brain parenchyma,[1,2]

The Choroid Plexus and Cerebrospinal Fluid. http://dx.doi.org/10.1016/B978-0-12-801740-1.00003-2

and is essential for the production of cerebrospinal fluid (CSF).[3] The structure of the CP is critical to its function. The central nervous system (CNS) is one of several organ systems considered to be an immune-privileged site, along with the eyes and testes. There are three principal CNS barriers that maintain immune-privilege in the CNS, yet allow specific immune cells to enter the CNS for immune surveillance, and enable immune cells to home to injured or diseased regions in the brain and spinal cord. These barriers are the CP, also referred to as the blood–choroid plexus barrier (BCPB), responsible for determining the specificity of the immune cells that enter the CSF and CNS,[1,2] the blood–brain barrier (BBB), an extensive microvessel complex throughout the brain parenchyma blocking toxins, microbial agents, and inflammatory cells from entering the brain, and the blood–leptomeningeal barrier (BLMB), an epithelial membrane covering the brain and spinal cord associated with the vasculature and CSF. The CP plays a prominent role in regulating cellular entry into the CNS.[3]

The CP unit (Fig. 3.1) contains an apical region composed of secretory epithelial cells responsible for CSF production. These cells contain numerous microvilli and cilia that are also responsible for circulating the CSF, which goes from the brain ventricles to the brain stem, into the subarachnoid space of the meninges, thereby covering the spinal cord and brain.[4] The region within the CP, beneath an apical layer of choroid epithelial cells, is composed of mesenechymal-like fibroblasts referred to as CP stromal cells.[4] The CP stroma is highly vascularized, containing capillaries that are unique to the brain. Unlike the vasculature comprising the BBB, the CP endothelial cells are thin, rest on a basal membrane, and exhibit diaphragmed fenestrations; and these blood vessels are relatively larger in diameter compared to microvessels in the brain,[4] These endothelial cells are highly permeable to fluid and solutes, including proteins.[5] Tracer studies using horseradish peroxidase or microperoxidase demonstrated the lack of barrier between the blood and extracellular space of the stroma.[6] These endothelial cells are not associated with astrocytes, and do not contain tight junction proteins. The blood vessels within the CP stroma are permeable to circulating inflammatory cells in normal and diseased states.[4] The number of pinocytotic vesicles in endothelial cells forming the BBB are very limited compared to the number of pinocytotic vesicles in CP endothelial cells.[5]

The barrier separating the CP stroma from the CSF is a wall of cuboidal epithelial cells formed by tight junction proteins similar

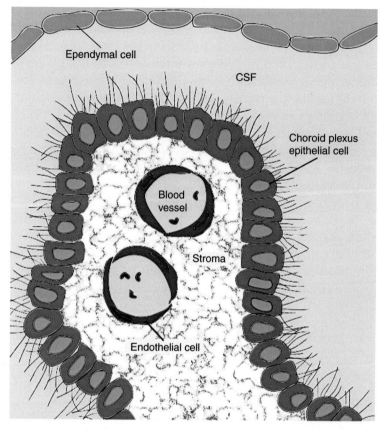

FIGURE 3.1 **Anatomy of the CP.** The CP is the organ localized to the lining of the brain ventricles, containing an apical region composed of secretory CP epithelial cells responsible for CSF production. Beneath the choroid plexus epithelial cells lie mesenechymal-like stromal cells. Relatively large capillary blood vessels are found within the CP stoma. These capillaries contain fenestrated endothelial cells, unique to the brain. Peripheral blood enters the CP stroma through the fenestrated endothelial vasculature. At this point, determination of the types of inflammatory cells entering the CSF is made by the CP epithelial cells. The expression of specific receptors and adhesion molecules on the CP epithelium will determine which types of cells will pass the barrier and enter the CSF. Specialized CP epithelial cells, ependymal cells, line the brain parenchyma. The CP is a gateway from the peripheral blood to the CSF and central nervous system. *Courtesy of Rachel Rosenstein-Sisson.*

to those found in the endothelial cell barrier in the BBB.[4] The structure of the CP is uniquely adapted for proper transmission of materials from the blood into the CSF. Ependymal cells line the surface of the brain parenchyma exposed to CSF (Fig. 3.1).

By contrast to the CP, the BBB is a structure composed of blood vessels lined with unique endothelial cells, whose properties include low numbers of pinocytotic vesicles, and high expression of several tight junction proteins, such as occludin, claudin, and ZO-1,2, which provide a physical barrier for leukocytes.[3] The BBB endothelial cells are tightly bound by the unique vascular unit consisting of pericytes, astrocytic end-feet, neuronal processes, as well as extracellular matrix secreted by endothelial cells.[4] This vascular unit provides an effective cellular and particulate barrier between the systemic blood supply and brain parenchyma in the normal, nondiseased state.

The BLMB is comprised of tight-junction endothelial cells, similar to the BBB, but is adjacent to the subarachnoid space. During an inflammatory event, adhesion molecules are upregulated on the endothelial cell surface, resulting in binding and entrance of specific inflammatory cells into the subarachnoidal space, which contains CSF, and ultimately into the adjacent CNS parenchyma.

INFLAMMATORY CELLS IN THE CHOROID PLEXUS

Although the CNS is an immune-privileged site, the CSF may contain inflammatory cells in healthy as well as in injured or diseased conditions.[1] The CP is the primary site of entry of these inflammatory cells.[1] In the normal brain, T cells are found in the CSF and in the meningeal membranes.[7] Immune cells enter the CNS through the vasculature associated with the CP and leptomeninges. In the CP, inflammatory cells traveling in the blood bind to the endothelial cell walls via specific adhesion molecules; subsequently, these cells migrate through the fenestrated vasculature of the CP into the CP stroma.[8] Leukocyte entry involves an initial tethering, and then rolling, followed by adherence to adhesion molecules expressed on activated endothelial cells. Leukocytes bind to surface endothelial cell adhesion molecules, such as ICAM-1, ICAM-2, VCAM-1.[1] Once bound, the leukocytes then protrude into the abluminal side of the vasculature. There are two distinct migration pathways for immune cells: paracellular diapedisis – immune cells go through endothelial cell junctions; and transcellular diapedisis – immune cells induce formation of pore-like structures in the endothelial cells permitting migration from the luminal to the abluminal side of the blood vessel,[1] leaving inflammatory cells in the CP stroma. The CP is a significant portal for macrophage entry into the CNS.

MACROPHAGES

The immune cells entering the CP include macrophages derived from circulating monocytes. Macrophage populations are generally classified into two populations based on their function following injury or inflammation of the CNS.[9] The initial incoming macrophage population is the proinflammatory cells, M1 macrophages, characterized as highly phagocytic, responsible for removing debris, and secreting proinflammatory cytokines that further enhance inflammation, particularly activating incoming T cells.[9] The principal cytokines produced by these M1 cells are TNF-α and IL-1β.[10] M1 macrophages can inflict tissue damage through the secretion of toxic factors such as ROS, proteases, and TNF-α. The second category of macrophages (M2) appear later during the adaptive immune response, and function as anti-inflammatory cells responsible for resolving the immune response and essentially wound healing. These M2 cells participate in tissue regeneration, angiogenesis, and matrix deposition for tissue remodeling. During the resolution of mechanical injury to the brain and spinal cord (SC) parenchyma, both macrophage populations are found at the injured site. However, macrophages enter the CNS parenchyma through different routes. M1 cells enter the CNS through the blood–leptomeninges barrier, in response to the secretion of CCL1 (MCP-1) by the meninges. M1 cells express the CCR2-receptor for ligand CCL1.[9] By contrast, M2 cells traffic through the CP. M2 cells breach the CP barrier by disrupting epithelial tight junctions, particularly tight junction protein ZO-1, and reducing occludin expression. Inflammation activates VCAM-1 on epithelial cells, enhancing M2 recruitment. Experiments blocking VCAM-1 expression resulted in reduced M2 presence in the CP.[10] The CP functions as an immune modulator, by expressing IDO (indoleamine-2,3-dioxygenase), which catalyzes tryptophan synthesis causing inhibition of T cell activity. IL-13 is the dominant cytokine in the CP and CSF, and inducer of the M2 population; TGF-β2 is highly expressed in the CSF and IL-10 is highly expressed in the CP. TGF-β2 and IL-10 are recognized immune suppressive cytokines and growth factors.[11] Also present in the CP are inducible Tregs (regulatory T cells), which produce the immunosuppressive agent adenosine.[12] The CP orchestrates trafficking of M2 macrophages into CSF; and the leptomeninges support entrance of M1 macrophages into the CSF during the acute phase. CP provides the site and supporting milieu for the maintenance and amplification of the anti-inflammatory M2 macrophage population.

Thus, the CP facilitates CNS immunosurveillance. Leukocytes extravagate across the endothelial cell boundary, and bind specifically to tight-junction CP epithelial cells. If appropriate receptors are present, these leukocytes then enter the CSF. The CP works as a selection organ for the appropriate macrophage population (M2) that secretes anti-inflammatory cytokines, which arrest proinflammatory microglial/macrophage function.

DENDRITIC CELLS

Dendritic cells (DC), prime activators of the immune response, are characterized by high expression of MHC II. Majority of DC are bone-marrow-derived, and effective antigen presenting cells (APC).[13,14] The CNS has resident DC in the CP, which express high levels of MHC II; these cells are referred to as epiplexus or Kolmer cells. Epiplexus cells are adherent to the apical region of CP epithelial cells, and have access to the flowing CSF.[1,15] These cells are therefore exposed to CNS antigens necessary for effective immune surveillance. Studies have shown that CNS-derived DC inhibit T cell proliferation, as compared to proinflammatory blood-borne DC.[16] Proinflammatory DC stimulate T cell activation by producing IFN-γ and IL-17, while CNS-DC do not produce these inflammatory cytokines.[16] The inhibitory DC, considered immature DC, preferentially bind to BBB endothelial cells; and are present in the CP and meningeal space.[13] The specific homing characteristic of the immature DC is the result of integrins, α4β1, expressed on DC; β1 is responsible for the initial binding of immature DC to the vasculature, α4 is responsible for firmness of binding.[17] Granulocyte-macrophage colony-stimulating factor (GM-CSF) is used to enhance immunization by activating the release of DC from the bone marrow and stimulating entry of monocytes into meninges and CSF.[18] DC are the rate limiting factor in neuroinflammation.[13] GM-CSF also expands the inhibitory DC making the function of this cytokine ambiguous.[16] However, the conclusion is that while DC are present in the CP, these cells function to decrease T cell activation and the immune response.

T LYMPHOCYTES

The T cells, predominant memory T cells, are found in the CSF of healthy as well as diseased individuals.[1] In the noninflammatory situation (healthy person) there are approximately 1–3000 leukocytes/mL in CSF. In the normal state, T cells in the brain parenchyma and CSF are nonactivated memory T cells. By contrast in CNS inflammatory disease, such as multiple sclerosis, proinflammatory Th17 cells penetrate the CP epithelial layer using the CCR6/CCL20 axis.[19–21] Activated Th17 overexpress the CCR6 chemokine receptor; CP epithelial cells constitutively produce CCL20, the ligand for CCR6. Thus, the CCR6/CCL20 is a mechanism whereby specific T cell subpopulations, as well as other immune cell subtypes (i.e., B cells, Treg), concentrate in the CP stroma. For immune surveillance, T cell entry from the stroma into the CSF is regulated by the expression of VCAM-1 and ICAM-1 ligands on CP epithelial cells; these ligands bind to the specific T cell surface receptors $\alpha4\beta1$ and LFA-1, respectively. The adhesion molecules enable only memory T cells and not proinflammatory T cells to enter the CSF.[19,22] There is a natural selection for these T cells in the CP, since the peripheral blood contains 70% neutrophils, and the CSF is predominantly T cells, with very few neutrophils. T cells expressing high CD73 are noted to be associated with CP epithelial cells but not the CP vasculature, and not in the parencyhma or leptomeningeal vessels of the brain or spinal cord. CD73 is an ectonuclease expressed on inducible Treg, and involved in the production of the immunosuppressive agent adenosine.[12] This demonstrates that the CP plays an active role in immune surveillance by enhancing immune suppression, rather than proinflammatory T cell function.[23]

CEREBROSPINAL FLUID

In addition to a physical barrier, the CP functions as a secretory organ, producing approximately 500 mL of CSF per day, enough to circulate the entire CNS three to four times (per day).[24] A principal function of the CSF is as a mechanical buffer to protect the brain from physical impact.[3] Epithelial cells of the CP secrete CSF at 0.4 mL/min, making this organ one of the most actively secreting epithelial cell populations in the body. Furthermore, these cells are long-lived, and have a low proliferation rate.[3] The contents found in the CSF fluid are formed by active secretion and not mere filtration of the blood plasma. This is exemplified by the data that the CSF is slightly hypertonic as compared with the plasma. There are several homeostatic mechanisms controlling ion transport across the epithelial layer in the CP.[3,25] Movement of ions across the epithelial layer is mediated by transporters and ion channels, and the polar distribution of proteins between the apical and basolateral membrane.

The CP–CSF system provides a rich source of lipids, hormones, microRNAs, and cytokines as well as selective inflammatory cells to the CNS. Relative to the serum, the CSF has low protein, glucose, K^+, and Ca^{++}, but high Na^+, Mg^{++}, and Cl^-. Energy-dependent mechanisms are necessary for the secretion of sodium and removal of potassium from CSF. Furthermore, CSF has been shown to stimulate proliferation of neuronal stem cells, which lie adjacent to the CP in the subventricular zone.[26] The concentration of glucose in the blood plasma may be as much as 10 times greater than that found in the CSF. Thus, CSF glucose is independent of plasma glucose. The influx of glucose into the brain across the BBB is mediated by the GLUT transports on the luminal side of the blood vessels. The CSF has two to three times lower levels of amino acids as compared to the plasma. The CSF regulates amino acid entry into the brain. The brain can synthesize such amino acids as glutamate, aspartate, glycine, and glutamine. Other amino acids are transported across the CP. The low concentration of amino acids in the CSF is due to active efflux.[3]

The CSF delivers thyroid hormone (T4) to the brain. Thyroid hormones bind to globulin and transthyretin (TTR). Since T4 is a lipid, this substance passes through plasma membranes by diffusion; T4 is higher in the CSF compared to plasma. The CP synthesizes TTR, and supplies this agent to the CNS. The TTR–T4 complex is taken up by endocytosis in the CP and T4 is taken up into the ependyma.[3]

The CP produces bioactive peptides and growth factors. Vasopressin (VP) is upregulated in the CSF compared to plasma. CP epithelial cells synthesize and secrete VP, which accumulates in the apical cells of the CP epithelium. CP epithelial cells have receptors for VP thereby making this an autocrine pathway.

The CSF contains several key growth factors including insulin-like growth factor-I and II (IGF-I and II),[24] which bind to the IGF-IR on glial progenitor cells, causing increased proliferation, as a result of acceleration of G1-S progression.[27] IGF-1 also functions as a neuroprotector by enhancing the clearance of brain amyloid β by modulating transport and production of amyloid carriers in the CP.[28] Brain-derived neurotrophic factor (BDNF) and IGF-I are upregulated in the CP and CSF in response to chronic stress.[29] IGF-II is present in the CP-epithelium, brain microvessels, and meninges. The CP epithelial cells express the IGF-I receptor and EGF receptor. The role of IGF-II in the brain is not well understood, and may function in cell survival and recovery. Transforming growth factor β1 (TGFβ1) is important for recovery from brain injury, and may be involved in calcium homeostasis. Common growth factors in the CSF are TGFβ1, hepatocyte growth factor (HGF), and bone morphogenic protein (BMP) 6 and 7, which play an important neuroprotective role for hippocampal neurons.[30]

The CP regulates metal ions in the CSF. Recent studies have shown that the CP is the site of regional iron homeostasis;[11] CP epithelial cells express toll-like receptor 4 (TLR4) under basal conditions. Therefore, when the CP is exposed to inflammatory agents, eg., lipopolysaccharides (LPS), these cells are activated. This activation stimulates the production of hepcidin, a mediator of iron export and degradation. Iron transfer into brain cells occurs at the CP and CSF. This transfer of iron is mediated by upregulation of transferin, transferin receptor type 1 and ferroportin, all of which are associated with CP epithelial cells. Inflammation causes CP epithelial cell upregulation of hepcidin and increased secretion of lipocalin 2, a protein that limits iron in CSF. During viral infection, iron is removed from the CSF; ependymal cells sequester iron from the CSF. Since microorganisms are sensitive to reduced iron availability, this regulation of iron is one mechanism whereby the CP and CSF reduce viral activity in order to decrease local immune activity. The CP also regulates lead. To be noted, high levels of lead result in reduction of TTR synthesis in the CP causing the disruption of thyroid uptake by the brain.[3]

CONCLUSIONS

The CP plays a crucial role in maintaining the homeostasis of the brain. Its role in CSF production, maintenance of the choroid plexus–CSF barrier, neuroimmune trafficking, and regulation of ions, peptides, and thyroid hormone are crucial to the wellbeing of the brain.

Acknowledgment

We wish to thank Rachel Rosenstein-Sisson for her skillful and creative artwork, and her valuable assistance in editing the manuscript.

References

1. Engelhardt B, Ransohoff RM. Capture, crawl, cross: the T cell code to breach the blood–brain barriers. *Trends Immunol*. 2012;33(12):579–589.
2. Shechter R, London A, Schwartz M. Orchestrated leukocyte recruitment to immune-privileged sites: absolute barriers versus educational gates. *Nat Rev Immunol*. 2013;13(3):206–218.
3. Redzic ZB, Segal MB. The structure of the choroid plexus and the physiology of the choroid plexus epithelium. *Adv Drug Deliv Rev*. 2004;56(12):1695–1716.
4. Ransohoff RM, Engelhardt B. The anatomical and cellular basis of immune surveillance in the central nervous system. *Nat Rev Immunol*. 2012;12(9):623–635.
5. Strazielle N, Ghersi-Egea JF. Physiology of blood–brain interfaces in relation to brain disposition of small compounds and macromolecules. *Mol Pharm*. 2013;10(5):1473–1491.
6. Brightman MW. The intracerebral movement of proteins injected into blood and cerebrospinal fluid of mice. *Prog Brain Res*. 1968;29:19–40.
7. Baruch K, Schwartz M. CNS-specific T cells shape brain function via the choroid plexus. *Brain Behav Immun*. 2013;34:11–16.
8. Wojcik E, Carrithers LM, Carrithers MD. Characterization of epithelial V-like antigen in human choroid plexus epithelial cells: potential role in CNS immune surveillance. *Neurosci Lett*. 2011;495(2):115–120.
9. Shechter R, Miller O, Yovel G, et al. Recruitment of beneficial M2 macrophages to injured spinal cord is orchestrated by remote brain choroid plexus. *Immunity*. 2013;38(3):555–569.
10. Tang Y, Le W. Differential roles of M1 and M2 microglia in neurodegenerative diseases. *Mol Neurobiol*. 2015;.
11. Marques F, Falcao AM, Sousa JC, et al. Altered iron metabolism is part of the choroid plexus response to peripheral inflammation. *Endocrinology*. 2009;150(6):2822–2828.
12. Whiteside TL. Regulatory T cell subsets in human cancer: are they regulating for or against tumor progression? *Cancer Immunol Immunother*. 2014;63(1):67–72.

13. Jain P, Coisne C, Enzmann G, Rottapel R, Engelhardt B. Alpha4beta1 integrin mediates the recruitment of immature dendritic cells across the blood–brain barrier during experimental autoimmune encephalomyelitis. *J Immunol.* 2010;184(12):7196–7206.

14. McMenamin PG. Distribution and phenotype of dendritic cells and resident tissue macrophages in the dura mater, leptomeninges, and choroid plexus of the rat brain as demonstrated in wholemount preparations. *J Comp Neurol.* 1999;405(4):553–562.

15. Anandasabapathy N, Victora GD, Meredith M, et al. Flt3L controls the development of radiosensitive dendritic cells in the meninges and choroid plexus of the steady-state mouse brain. *J Exp Med.* 2011;208(8):1695–1705.

16. Hesske L, Vincenzetti C, Heikenwalder M, et al. Induction of inhibitory central nervous system-derived and stimulatory blood-derived dendritic cells suggests a dual role for granulocyte-macrophage colony-stimulating factor in central nervous system inflammation. *Brain.* 2010;133(Pt 6):1637–1654.

17. Zhan Y, Xu Y, Lew AM. The regulation of the development and function of dendritic cell subsets by GM-CSF: more than a hematopoietic growth factor. *Mol Immunol.* 2012;52(1):30–37.

18. Kingston D, Schmid MA, Onai N, Obata-Onai A, Baumjohann D, Manz MG. The concerted action of GM-CSF and Flt3-ligand on *in vivo* dendritic cell homeostasis. *Blood.* 2009;114(4):835–843.

19. Sallusto F, Impellizzieri D, Basso C, et al. T-cell trafficking in the central nervous system. *Immunol Rev.* 2012;248(1):216–227.

20. Qi W, Huang X, Wang J. Correlation between Th17 cells and tumor microenvironment. *Cell Immunol.* 2013;285(1–2):18–22.

21. Schwartz M, Baruch K. The resolution of neuroinflammation in neurodegeneration: leukocyte recruitment via the choroid plexus. *EMBO J.* 2014;33(1):7–22.

22. Kunis G, Baruch K, Rosenzweig N, et al. IFN-gamma-dependent activation of the brain's choroid plexus for CNS immune surveillance and repair. *Brain.* 2013;136(Pt 11):3427–3440.

23. Kivisakk P, Mahad DJ, Callahan MK, et al. Human cerebrospinal fluid central memory CD4+ T cells: evidence for trafficking through choroid plexus and meninges via P-selectin. *Proc Natl Acad Sci USA.* 2003;100(14):8389–8394.

24. Lehtinen MK, Bjornsson CS, Dymecki SM, Gilbertson RJ, Holtzman DM, Monuki ES. The choroid plexus and cerebrospinal fluid: emerging roles in development, disease, and therapy. *J Neurosci.* 2013;33(45):17553–17559.

25. Pardridge WM. Drug transport across the blood–brain barrier. *J Cereb Blood Flow Metab.* 2012;32(11):1959–1972.

26. Lehtinen MK, Walsh CA. Neurogenesis at the brain–cerebrospinal fluid interface. *Annu Rev Cell Dev Biol.* 2011;27:653–679.

27. Yeh C, Li A, Chuang JZ, Saito M, Caceres A, Sung CH. IGF-1 activates a cilium-localized noncanonical Gbetagamma signaling pathway that regulates cell-cycle progression. *Dev Cell.* 2013;26(4):358–368.

28. Carro E, Spuch C, Trejo JL, Antequera D, Torres-Aleman I. Choroid plexus megalin is involved in neuroprotection by serum insulin-like growth factor I. *J Neurosci.* 2005;25(47):10884–10893.

29. Schwartz M, Baruch K. Vaccine for the mind: immunity against self at the choroid plexus for erasing biochemical consequences of stressful episodes. *Hum Vaccin Immunother*. 2012;8(10):1465–1468.
30. Krieglstein K, Strelau J, Schober A, Sullivan A, Unsicker K. TGF-beta and the regulation of neuron survival and death. *J Physiol Paris*. 2002;96(1–2): 25–30.

Toward an Artificial Choroid Plexus, Concept and Clinical Implications

Thomas Brinker, John Morrison

Department of Neurosurgery, Warren Alpert Medical School, Brown University, Rhode Island Hospital, Providence, RI, USA

INTRODUCTION

The understanding of cerebral spinal fluid (CSF) physiology is continuously evolving. Classical, more mechanistic, notion of CSF circulation from choroid plexus (CP) to arachnoid granulations is giving way to a more complex model with the blood–brain

The Choroid Plexus and Cerebrospinal Fluid. http://dx.doi.org/10.1016/B978-0-12-801740-1.00004-4

barrier (BBB) serving as the site of CSF production and absorption. Furthermore, the role of the CP as the major site of CSF production and absorption is challenged. However, the role of CP as a source of biologically active compounds, that is peptides, is becoming more evident. The CP-secreted peptides may be involved in the regulation of brain homeostasis, especially in CNS diseases. We will discuss that the novel understanding of CSF physiology does not depend upon the function of CP as a major source of CSF production. We will then review evidence that indicates an important role of CP as a source of peptides secreted into the CSF under physiological and pathophysiological conditions. Also, pharmacological factors are discussed, such as whether the CSF space may serve as a compartment and system for the distribution of peptides throughout the brain. Finally, we analyze our previous work with encapsulated stem cells transplanted into the CSF space and the concept of an artificial CP is outlined.

THE CP IS THE MAJOR SITE FOR CSF FORMATION, REALLY?

CP tissue is floating within the cerebrospinal fluid of the lateral, third, and fourth ventricles. The tissue features numerous villi, which are each well perfused, harboring capillaries with fenestrated endothelium. A single layer of cuboidal epithelium then covers each of these vessels. The cellular anatomy forming the blood–CSF barrier is characterized by tight junctions at the apical end of the choroid epithelial cells rather than at the capillary endothelium.[1,2] Although it was traditionally recognized that the ependyma and the brain parenchyma are supplementary CSF sources, CP has been regarded as the major site of CSF production (reviewed by McComb[2]). Interestingly, this notion is mainly based upon the historical canine experiments of Dandy who occluded the foramen of Monroe and performed a choroid plexectomy of one lateral ventricle. Since the collapse of the ventricle without CP and the dilatation of the contralateral ventricle were found,[3] it was concluded: "From these experiments we have the absolute proof that cerebrospinal fluid is formed from the choroid plexus. Simultaneously it is proven that the ependyma lining the ventricles is not concerned in the production of cerebrospinal fluid".[4] According to a previous review,[5] the experiments of Dandy were based upon observations from only a single dog. Furthermore, these experiments could not be reproduced by others[6,7] and no hydrocephalus was reported following the occlusion of the aqueduct cerebri.[8] There were two other sets of experiments that were thought to be "crucial" in support of Dandy's thesis.[5] First, the hematocrit of the CP blood was found to be 1.15-times greater than of that of the systemic arterial blood. From this value, and the estimated arterial blood flow through the CP, a CSF secretion rate was calculated that came very close to the total estimated rate of CSF absorption.[9] Second, these findings were substantiated by assessment of the CSF production rate in the isolated and extracorporally perfused CP.[10-13] These experiments were criticized because of inherent large errors in the complex experimental technique.[1,2,5] Furthermore, it was shown that some CSF must come from a source other than the CP, presumably the brain tissue itself, using radioactive-water experiments.[14-16] Also, perfusion studies performed on isolated regions of the ependymal surface revealed that nearly 30% of the total CSF production may be produced by the ependyma.[17] Even higher fractions of ependymal fluid secretion were found while investigating spinal cord ependyma.[18] Again, these experiments were criticized because of the, "drastic experimental procedures" required. It was concluded, "it may be wise to reserve final judgment on this question".[1]

However, recent molecular biological insights into the function of the BBB confirm the earlier notion [19] that cerebral capillaries are active producers of CSF. The discovery of water transporters located at the end-feet processes of astrocytes has decisively improved our understanding of the physiology of the BBB and has led to the notion that large water fluxes continuously take place between different fluid compartments of the brain, that is the blood, CSF, and interstitial fluid (ISF) (reviewed by Tait[20]).[21,22]

However, there remains a great uncertainty about the rate of fluid formation. The formation rate at the BBB was calculated from the absorption rate of tracers, which were injected into the brain parenchyma. It was found to be considerably lower than the formation rate at CP.[23] Interestingly, such extensive water movements were indicated by earlier radiotracer experiments. For example, following the intravenous injection of deuterium, a rapid distribution throughout all brain compartments was reported as early as 1952.[24] This data demonstrated water fluxes that greatly exceeded the contemporary estimated rates of CSF and ISF flow. Recently, the original data on the deuterium half-life in different brain compartments has been used to calculate the respective CSF fluxes by applying magnetic resonance imaging (MRI)-based volume assessments of the ventricles, the subarachnoid space, and the spinal CSF spaces. As a result, CSF fluxes of more than 22 mL/min and a CSF turnover rate of more than 140-times a day were calculated, values that greatly exceed the CP production.[25,26] Considering these discrepancies, a recent review concluded, "the working hypothesis that the BBB is a fluid generator, although attractive, needs substantiation".[27]

However, the same argument could be advanced regarding the CSF formation at the CP, since all techniques used to calculate the formation rate are a matter of debate. In 1931, Masserman calculated the rate of CSF formation in patients by measuring the time needed for CSF pressure to return to its initial level following drainage of a standard volume of CSF by lumbar puncture. After drainage of 20–35 mL of CSF, pressure was restored at a rate of about 0.32 mL/min.[28] The validity of results obtained in this way was criticized because the Masserman technique assumes that neither formation nor absorption rates are changed by alterations in pressure. However, absorption of CSF varies greatly with changes in intracranial pressure.[29,30] Modifications of the Masserman technique applied sophisticated infusion and drainage protocols, which recorded and controlled the CSF pressure during the measurement period (see, e.g., Ekstedt[31]). Despite numerous research efforts, more sophisticated drainage protocols did not yield CSF formation rates that differed from earlier work. The ventriculo-cisternal perfusion ("Pappenheimer") technique represents a more quantitative approach for the assessment of ventricular CSF formation. Inulin, or another macromolecule, which is not absorbed in the ventricles, is infused at a constant rate into the cerebral ventricles. CSF formation is calculated from the measurement of the extraventricular (cisternal or spinal) CSF concentration of inulin. It is assumed that any dilution of inulin results from the admixture of freshly formed CSF. Clinical measurements were performed in brain tumor patients who received ventricular catheters for chemotherapy purposes. In patients (9–61 years old) the average flow rate was 0.37 mL/min.[32] These results were confirmed in brain tumor children.[33] Though more precise, the ventriculo-cisternal or ventriculo-lumbar perfusion techniques yielded results remarkably close to those assessed by the Masserman technique.[2] Findings from the Masserman and the Pappenheimer techniques were supported by neuroradiological investigations applying serial computed tomography (CT) scan to assess the ventricular washout of metrizamide, a water-soluble contrast media. The rate of right lateral ventricular CSF formation ranged from 0.0622 mL/min to 0.103 mL/min.[34,35] However, as has been discussed in detail by others[36,37], the calculation of choroidal CSF production, by means of the Pappenheimer technique, depends on the assumption that CP is the only intraventricular source of CSF and that no CSF inflow or outflow via the periventricular tissue occurs. Unfortunately, according to radiotracer studies[24,38] such assumptions are questionable at best.[39] Thus, until today, the assessment of the CSF formation and absorption rate remained a matter of debate, "a method for measuring CSF formation and absorption rates less invasive than the Pappenheimer method (ventriculo-cisternal perfusion) and more reliable than the Masserman method is sorely needed".[30] Interestingly, MRI flow measurements revealed conflicting results too: phase contrast MRI may provide quantitative blood-flow velocity along the aqueduct in humans.[40–42] Advancing phase contrast MRI, the cine phase contrast technique yields quantitative flow information by synchronizing the acquisition of the images to the cardiac cycle.[43,44] Applying these techniques, the normal aqueduct flow has been measured many times in adults. Flow rates were reported ranging from 0.304 mL/min to 1.2 mL/min.[45–49] Based upon this data, the average normal flow in healthy adults was suggested to be 0.77 mL/min in craniocaudal direction.[26] Hence, CSF flow measured by MRI exceeds the customarily assumed choroidal CSF production rate by a factor of two. Technical limitations of the MRI flow measurements must be considered before interpreting these MRI data that are not congruent with the traditional understanding of CSF physiology. Thus, it was pointed out that the evaluation of the flow void is subjective and highly dependent on the acquisition parameters used, as well as on the technical characteristics of the MRI systems (e.g., gradient strength).[41]

Against this background, one may conclude that the customary understanding of the CP to be a major site of CSF production may be questionable. This view is supported by clinical observations following surgical removal of CP for the treatment of hydrocephalus. Choroidectomy was widely performed in the 1950s and eventually not pursued since it was, in the long term, not sufficient for the treatment of hydrocephalus.[7] Explaining the disappointing therapeutic results, it was argued that it is not possible to remove the CP tissue completely, since the tissue in the fourth ventricle is not surgically accessible (Milhorat[7]; see also Milhorat[26]).

However, clinical findings following choroidectomy suggest alternative CSF production sites, that is BBB may compensate for the lost production at the CP. However, one may also argue that the CP is not the only major site of CSF formation and that the production at the BBB is quantitatively more important.[39] At this point, one has to acknowledge the renaissance of choroidectomy in combination with a third ventriculostomy in hydrocephalic children (Stone and Warf[50]; Warf[51]). The existing clinical data indicates the therapeutic value, though no prospective randomized trial including long-term follow up data is available thus far.

CP AS A SITE FOR SECRETION OF BIOLOGICALLY ACTIVE COMPOUNDS

Although current research puts the significance of CP for the production of CSF in the rear, its role for development and "maintenance" of brain function is highlighted. Especially in brain diseases, CP serves as a site for the invasion of blood borne cells, which modulate, for example, neuroinflammation (see recent review by Szmydynger-Chodobska et al.[52]). The choroidal secretion of biologically-active compounds, that is peptides and proteins, is possibly more important, even under physiological conditions. Peptides are ubiquitous in the brain, serving as small-cell-signaling protein molecules and playing a pivotal role in intercellular communication. Different functions are allocated to neuropeptides for interneuronal signaling. Specifically, neurotrophins are responsible for synaptogenesis, anti-apoptotic effects, and neuroregeneration; endocrine factors regulate and coordinate brain and systemic metabolic homeostasis; and chemotactic and cytokine factors modulate brain inflammation. Additionally, peptides are heavily involved in the interaction of astrocytes, microglia, pericytes, and neurons.

Since it has been shown that peptides play an important role, not only in the healthy brain, but also in acute and chronic brain disorders, research has aimed to harness their putative therapeutic potential. Therapeutic effects of peptides have been reported in a myriad of experimental studies. Excellent reviews summarize the efficacy of neuropeptides in *chronic* neurodegenerative disorders[53-55] and also in normal ageing.[56] Also, therapeutic potential of peptides for the treatment of acute brain disorders has been recognized.[57-59] Since excellent reviews are available pointing to the CP as a source of such peptides (see, e.g., Refs 59–63), and other chapters of this book deal in detail with this, we refer to those sources. Instead, this chapter will take it for granted that the CP acts as an organ not only secreting CSF but possibly more important, biologically active compounds, that is neuropeptides, hormones, and vitamins.

PHARMACOLOGICAL CONSIDERATIONS FOR THE DISTRIBUTION OF PEPTIDES WITHIN THE CSF SPACE

At this point we want to discuss pharmacological issues, that is how biologically active compounds secreted at the CP could be distributed throughout the brain. This is an important issue since in the past, the idea has been strictly refused that CSF space could serve as a medium for the administration of therapeutically active substances to the CNS (see, e.g., Refs 64–67). The potential of drug delivery via the CSF has been considered to be low because of (1) a very limited diffusion of molecules through the brain parenchyma and (2) a rapid wash-out of drugs from cerebrospinal fluid into the blood. Accordingly, a previous review assumes that CP-secreted molecules just target periventricular sites of the brain.[68] It was further reported that following the administration of drugs into the CSF space, the penetration into periventricular sites of the brain decreases exponentially with the distance from the CSF surface.[69] Thus, it is unclear whether agents secreted by the CP may effectively target brain structures other than those of the periventricular sites.

The extracellular space (ECS) of the brain has a narrow and tortuous structural architecture. The width of the ECS in living tissue is estimated to range only between 38 nm and 64 nm.[70] As a result, diffusion of peptides, especially when larger than 10 kDa, is slow, and diffusion distances are limited.[71] This was also shown *in vivo* with intraparenchymal implantation of controlled release polymers containing neurotrophic growth factor. Analysis of sequential sections on the autoradiograph confirmed that ^{125}I-labeled NGF was transported only 2–3 mm from the polymer in any direction.[72] Different polymeric delivery systems did not achieve a deeper penetration into the brain parenchyma.[73]

The rapid bulk clearance rate and venous absorption of CSF were suggested as mechanisms preventing significant diffusion of drugs into the brain following intrathecal injection.[74] Similarly, binding protein components of the CSF and ependymal, pial, and glial tissues may hamper tissue penetration.[75] Such pharmacokinetic limitations may explain the notion that administration of agents into the CSF consistently requires higher doses to elicit effects similar to those obtained with intraparenchymal delivery.[71] For example, the effect of NGF on mRNA levels for choline acetyltransferase (ChAT) in striatal cholinergic cells in rats requires a daily dose of only 50 ng with intracerebral administration, but 4.5 μg with the intraventricular route.[76]

However, tracer studies depicted that even larger molecules may be very well distributed throughout the brain following intraventricular administration: 4 h after intraventricular or intracisternal (ICV) infusion of ^{14}C-labeled inulin, autoradiography showed in rats the entire brain labeled, irrespective of the site of the tracer application.[77] The distribution of radioiodinated recombinant human nerve growth factor ([^{125}I]rhNGF) was evaluated in adult cynomolgus monkeys following unilateral ICV administration. Autoradiography performed 24 h after the infusion showed specific radiolabel not only bilaterally throughout the basal forebrain, but also in the superficial ventral cortex.[78] Notably, the intraventricular route has been shown to be superior to the intravenous one for the delivery of lentiviral-mediated gene therapy.[79] Another study described a lentiviral-based system, administered via the ICV route, for the delivery peptides or proteins into the cerebrospinal fluid.[80]

Following the intraventricular administration, ^{14}C-sucrose was quickly transported to the basal cisterns and penetrated into the perivascular spaces (PVS) where it persisted for more than 3.5 h.[81] This finding suggests that, following the administration into the CSF, the brain-wide distribution of agents may depend on the penetration of the molecules into the PVS. Anatomically, a PVS extends along arteries and veins penetrating from the subarachnoid space into the brain parenchyma. The PVS is obliterated at the capillary level and, furthermore, it is separated from the subarachnoid space by a

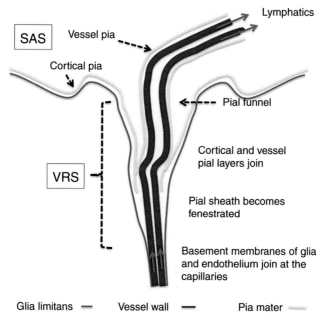

Lymphatics

SAS Vessel pia

Cortical pia

Pial funnel

Cortical and vessel
pial layers join

VRS

Pial sheath becomes
fenestrated

Basement membranes of glia
and endothelium join at the
capillaries

Glia limitans — Vessel wall — Pia mater

FIGURE 4.1 **Morphology of the PVS.** Delineated by basal membranes of glia, pia, and endothelium, the PVS depicts the space surrounding vessels penetrating into the parenchyma. The PVS is obliterated at the capillaries where the basement membranes of glia and endothelium join. The complex pial architecture may be understood as an invagination of both cortical and vessel pia into the PVS. The pial funnel is not a regular finding. The pial sheath around arteries extends into the PVS, but becomes more fenestrated and eventually disappears at the precapillary section of the vessel. Interstitial fluid may drain by way of an intramural pathway along the basement membranes of capillaries and arterioles into the lymphatic at the base of the skull (green arrows). VRS, Virchow–Robin space; SAS, subarachnoid space. *Reprinted with permission from Ref. [83].*

pial membrane. Therefore, rebutting earlier assumptions,[82] there is no free circulation of CSF along the PVS. However, since the pial membrane consists of just a single layer of loosely connected pial cells without tight junctions, it constitutes just a diffusional barrier. Thus, the PVS is not anatomically, but functionally connected to the (subarachnoid) CSF space (Brinker et al.,[83]; Fig. 4.1). Accordingly, the protein tracer horse radish peroxide (HRP), following its administration into the CSF space, has been shown to deeply penetrate into the brain in dogs and cats. Indeed, a "paravascular" distribution was depicted along the PVS of vessels.[84] Similarly, following the administration into CSF in rats, labeled IGF-1 was detected predominantly in the pia mater, PVS, and subcortical white matter tracts 0.5 h after administration.[85] Very recently, the group of Nedergaard demonstrated, *in vivo*, the distribution of tracer molecules of different molecular weight along the PVS. Additionally, these experiments demonstrated that smaller molecular weight substances could diffuse from the PVS into the extracellular brain space.[86] It has been previously shown that the movement of tracers along the PVS depends on the transmission from the pulsations of the cerebral arteries to the microvasculature (Rennels et al.,[84]; reviewed by Rennels et al.[87]). The recent studies of Nedergaard revealed that, in addition, the perivascular distribution of tracers along the PVS depends on the activity of water transporters at the BBB: the penetration of cisternally injected tracer into the PVS was significantly decreased in mice lacking aquaporin-4.[86] Such findings support the earlier notion that the perivascular route may be a physiologically important route for the distribution of CSF-borne therapeutic molecules throughout the brain.[88,89] It should be noted that, in humans, electron microscopy shows the PVS to be collapsed.[90] This finding led to the notion that in humans the PVS represents an optional, rather than a physiologically open space (discussed by Brinker et al.[83]). Though, contrary to this notion, studies in rodents have consistently demonstrated the PVS filled with fluid, electron-microscopic-dense material,[91] macrophages, and other blood-borne inflammatory cells.[92,93] Different fixation procedures may explain this discrepancy; rodent brains undergo intravital perfusion fixation, while the studies in man have to rely on extracorporally fixed specimens.[83] Therefore, we suggest that, in humans too, the PVS may be understood as an extension of the subarachnoid CSF space penetrating deeply into the brain. This notion is supported by MRI findings, which clearly demonstrate PVS in humans.[94,95] Figure 4.2 illustrates possible fluid movements along the PVS.

Against this background, one may conclude that the PVS actually serves as a physiologically important route for the brain-wide distribution of peptides injected or secreted into the CSF space. Circulation of CSF-borne agents along the PVS could prove to be more important, since the capillary network of the brain is extraordinarily dense; every neuron in the brain is perfused by its own blood vessel and the distance between vessel and neuron is not more than 20 μm. Such a short distance makes diffusion nearly instantaneous.[67] However, though CSF-borne agents may be distributed along the PVS into the brain, the molecules may be rapidly cleared via the BBB

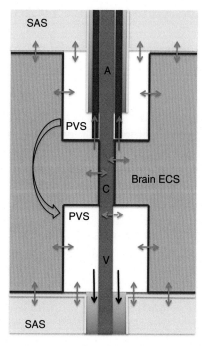

FIGURE 4.2 Fluid movements (blue arrows) along the PVS. The complex anatomical structure of the PVS allows a bidirectional fluid exchange between the PVS and both the brain ECS and the subarachnoidal CSF space. Therefore, molecules injected into the CSF space may enter the PVS, spread along the capillaries and diffuse into the ECS of the brain (see text). Glial (blue lines) and pial (yellow lines) cell membranes enclose the PVS and represent diffusional barriers. It is unclear, whether the subpial PVS around arteries and veins (light blue) serve as additional pathways for fluid exchange. Also, the proposed glymphatic pathway[86] connecting the arterial and venous PVS (black arrows) is still a matter of debate. A, artery; V, vein; C, capillary; SAS, subarachnoid space. *Reprinted with permission from Ref. [83].*

before penetrating into the neuropil. Such clearance from the brain is mainly based on membrane transport mechanisms; the rate of clearance depends on the specific properties of the molecules and the involved barrier transporters.[67] Nevertheless, in considering the PVS as a pathway for the distribution of molecules administered or secreted into the CSF, therapeutic effects at the perivascular site are reasonable.[83] For example, molecules secreted by the CP could modulate the BBB (or more precisely, the neurovascular unit) and additionally target the cellular components of the parenchyma throughout the entire brain.

TOWARD AN ARTIFICIAL CHOROID PLEXUS

Considering the therapeutic efficacy of neuropeptides and their secretion by the CP, it was quickly recognized that transplantation of CP tissue might exhibit neuroprotective, anti-inflammatory, or regenerative effects in the diseased brain (see, e.g., Emerich and Borlongan; Borlongan et al.[59,96]). It was also recognized that the transplantation of nonautologous CP cells caused a local tissue response, that is, gliosis, and more importantly the transplanted cells did not survive. The same experiments revealed that encapsulation of the cells prevented fibrosis and cell death.[97]

Encapsulation was originally introduced to achieve immune isolation allowing for allogenic or xenogenic cell transplantation. Semipermeable hollow fibers[98] as well as spherical polymeric microcapsules[99] protect cells transplanted into the brain against the immunological host-versus-graft response. Since the semipermeable capsule membrane permits the free passage of nutrients, oxygen, and, indeed, smaller molecules, the cells are preserved within the capsules, and can produce and deliver therapeutic peptides to the brain. The concept has been coined as "encapsulated cell biodelivery".[100] Figure 4.3 depicts the concept of cell encapsulation.

Encapsulation achieves the long-time survival of cells and thus the long-lasting and continuous release of biologically active

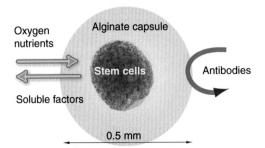

FIGURE 4.3 **Illustration of cell encapsulation technology.** Cells are embedded within a polymeric matrix, which is surrounded by an outer cell-free capsule. Smaller molecules diffuse through the capsule, while at the same time, larger proteins and cells are excluded. In this way the encapsulated cells are protected against the immune response of the host and continue to secrete biologically active compounds. Hundreds to thousands of those cell capsules may be transplanted into the CSF space and may achieve pharmacologically effective levels of secreted peptides.[101] *Reprinted with permission from Ref. [102].*

compounds. Viability of encapsulated-CP-epithelial cells for up to 7 months was demonstrated *in vitro*.[103] BHK cells, retrieved 1 year after transplantation into the ventricles of rats, continued to produce human neuronal growth factor (hNGF).[104] Such a proof of principle has also been shown in a clinical setting with both hollow fiber[105] and microcapsule implants (Brinker et al., unpublished data, ClinicalTrials.gov, identifier: NCT01298830).

Encapsulated Choroid Plexus Cells

Considering the *in vitro* evidence showing the neuroprotective potential of CP cells, the pioneering experiments of Borlongan showed in 2004 that encapsulated CP cells exhibited robust neuroprotective effects in a rodent model of acute stroke.[106] Some years later, the same group reviewed the literature and outlined a roadmap for the anticipated clinical translation of encapsulated-CP epithelium (CPE).[59] While there was great optimism regarding the therapeutic potential of encapsulated CP cell transplantation into the CSF space, the review listed important obstacles for a clinical translation. The authors discussed that a clinical translation needs a better understanding of the mechanisms of how encapsulated CP transplantation works in brain diseases, that is, there is a need to identify those secretory products and their relation to observed benefits. Also, the lack of highly purified epithelial cell line was recognized as being disadvantageous for a clinical translation. Finally, the limited availability of human CP tissue and the issues of the alternative, that is, to use xenogenic tissue, were discussed. Eventually, the authors identified the most urgent step as the need to develop techniques to generate homologous-CP cell populations, which could be proliferated and cryopreserved.[59] Unfortunately, it seems that those issues identified as obstacles for a clinical translation of CP transplantation remain imminent until today, though there is considerable progress. Thus, today, the best techniques available to study transport processes or polarized secretion by the CP are to create monolayer cultures of CPE.[107] Nonetheless, there remain major difficulties. Thus, it was criticized that "only limited attempts to culture these cells have been published, and they mainly include data from neoplastic CPE".[107] Representing a potential step forward, a CP cell line is commercially available. This in itself is however, controversial, as its origin remains obscure (i.e., whether it was proliferated from human CP taken postmortem or from fetuses after abortions).[107] Of note, currently in the United States and the European Union, cell lines derived either from malignant tumors or those with unknown origin do not meet regulatory needs for the approval of a clinical trial. Against this background (i.e., the lack of well characterized, bankable CPE cell lines), we suggest the use of genetically engineered mesenchymal stem cells to study the benefits of specific CP factors in health and disease. The idea is actually based on the earlier notion of Borlongan[106] that nonchoroidal cell lines that have been genetically modified to produce a specific therapeutic protein might be a valuable alternative to the implantation of CP tissue.

Encapsulated Mesenchymal Stem Cells

Today, there is consensus that stem cells, transplanted for the treatment of neurological disorders, exhibit therapeutic effects mostly by the secretion of soluble factors. The original notion that transplanted cells replace damaged neural tissue has shifted into the background.[108–112] Mesenchymal stem cells (MSCs) are the prototype of a cell line acting mainly by the secretion of peptides. Because of their abundant secretory activity, native MSCs were proposed "as site-regulated 'drugstores' *in vivo*".[113] For the sake of safety, mesenchymal stem cell lines and their cognates are currently favorites for cell transplantation, especially since highly standardized and well characterized bankable allogeneic cell lines are requested for clinical translation.[114–116] The MSCs may be genetically engineered to overexpress factors and used as carriers for the delivery of specific peptides. Such *ex vivo* gene therapy is considered to possess an extraordinary potential for the treatment of neurological disorders.[117–121] Assuming that the biological function of the CP depends on the secretion of biologically active compounds, it is obvious that the intraventricular transplantation of cell capsules containing genetically engineered MSC that could serve as an experimental model for the investigation of the CP function under both healthy and pathological conditions. This notion is confirmed by preclinical studies of our group showing the therapeutic potential of encapsulated MSCs that were genetically engineered to secret the neuroprotective gut hormone glucagon-like peptide-1 (GLP-1) in models of traumatic brain injury,[101] Alzheimer's disease,[122] and amyotrophic lateral sclerosis.[123] However, it should be noted that a clinical phase 1 trial with implantation of such engineered MSC capsules into the hemorrhagic brain in patients suffering from intracerebral hemorrhage showed a limited survival rate (0–30%) of the transplanted cells (unpublished data, ClinicalTrials.gov, identifier: NCT01298830, see also Heile and Brinker[124]). Though, this trial and another one investigating hollow fiber encapsulated cells in Alzheimer disease patients[125] indicate the safety of encapsulated cell biodelivery of peptides, and are therefore important steps toward a clinical translation.

Hence, regarding anticipated clinical trials investigating the intraventricular transplantation of secretory active cells, one has to consider findings of previous clinical trials describing unacceptable adverse effects after intraventricular peptide administration (reviewed by Emborg and Kordower[126] and Thorne and Frey[71]). The side effects were factor specific, for example after NGF administration, pain was the predominant symptom and after GDNF administration nausea and behavioral changes were observed. Significant side effects causing a permanent cellular pathology were reported from experimental studies. The ICV administration of NGF caused axonal sprouting and Schwann cell hyperplasia on the ependymal or arachnoid surface[127] and hyperplastic changes within the leptomeninges of the rat and monkey.[128] FGF-2 resulted in periventricular astrogliosis.[129] FGF-2[130] and bFGF[131] were reported to induce symptomatic hydrocephalus in rats. To address the possibility of serious adverse effects, it has been pointed out that cell transplants must be retrievable.[100] Concepts for the retrievability of encapsulated cells have been developed for hollow fiber[100] and microcapsules[102] cell implants.

CONCLUSIONS

First, the intraventricular transplantation of encapsulated genetically engineered stem cells of nonchoroidal origin may serve as a valuable experimental tool to demonstrate the role of the CP as an organ secreting biologically active compounds into the CSF space. Next, transplantation of genetically modified cell lines would allow studying the effects of specific factors of interest in the developmental, healthy, diseased, and aging brain. Furthermore, the therapeutic potential of factors could be investigated in a variety of brain disease models.

Regarding a clinical translation, that is, the transplantation of an "artificial CP" for therapeutic purposes, future research has to systematically investigate pharmacokinetic issues. The possibility of rapid membrane transporter clearance of CSF-borne compounds at the BBB, the limited diffusion of larger molecules through the brain parenchyma, or the binding of compounds by ependymal cell layers all require further understanding. Regarding any clinical translation, it is most important to identify possible safety issues since several previous clinical trials had to be terminated because of severe side effects following the administration of neuropeptides into the CSF. Encapsulated genetically modified cells with a known secretory profile may serve as an experimental paradigm to investigate such pharmacological issues. Of course, the therapeutic prospects of an artificial CP are exciting, not only for treatment of brain diseases, but also to compensate for the diminishing secretory potential of the CP seen in normal ageing.

References

1. Cserr HF. Physiology of the choroid plexus. *Physiol Rev.* 1971;51(2):273–311.
2. McComb JG. Recent research into the nature of cerebrospinal fluid formation and absorption. *J Neurosurg.* 1983;59(3):369–383.
3. Dandy WE, Blackfan KD. An experimental and clinical study of internal hydrocephalus. *JAMA.* 1913;61(25):2216–2217.
4. Dandy WE. Experimental hydrocephalus. *Ann Surg.* 1919;70(2):129–142.
5. Milhorat TH. The third circulation revisited. *J Neurosurg.* 1975;42(6):628–645.
6. Hassin GB, Oldberg E, Tinsley M. Changes in the brain in plexectomized dogs with comments on the cerebrospinal fluid. *Arch Neurol Psychiatry.* 1937;38:1224–1239.
7. Milhorat TH. Failure of choroid plexectomy as treatment for hydrocephalus. *Surg Gynecol Obstet.* 1974;139(4):505–508.
8. Oreskovic D, Klarica M, Vukic M. The formation and circulation of cerebrospinal fluid inside the cat brain ventricles: a fact or an illusion? *Neurosci Lett.* 2002;327(2):103–106.
9. Welch K. Secretion of cerebrospinal fluid by choroid plexus of the rabbit. *Am J Physiol.* 1963;205:617–624.
10. Rougemont de, Ames III A, Nesbett FB, Hofmann HF. Fluid formed by choroid plexus; a technique for its collection and a comparison of its electrolyte composition with serum and cisternal fluids. *J Neurophysiol.* 1960;23:485–495.
11. Ames JIII A, Sakanoue M, Endo S. Na, K, Ca, Mg, and Cl concentrations in choroid plexus fluid and cisternal fluid compared with plasma ultrafiltrate. *J Neurophysiol.* 1964;27:672–681.
12. Pollay M, Stevens A, Estrada E, Kaplan R. Extracorporeal perfusion of choroid plexus. *J Appl Physiol.* 1972;32(5):612–617.

13. Pollay M. Formation of cerebrospinal fluid. Relation of studies of isolated choroid plexus to the standing gradient hypothesis. *J Neurosurg.* 1975;42(6): 665–673.

14. Weed LH. The development of the cerebrospinal spaces in pig and in man. *Contrib Embryol Carnegie Inst.* 1917;5:1–116.

15. Bering Jr EA. Cerebrospinal fluid production and its relationship to cerebral metabolism and cerebral blood flow. *Am J Physiol.* 1959;197:825–828.

16. Bering Jr EA, Sato O. Hydrocephalus: changes in formation and absorption of cerebrospinal fluid within the cerebral ventricles. *J Neurosurg.* 1963;20: 1050–1063.

17. Pollay M, Curl F. Secretion of cerebrospinal fluid by the ventricular ependyma of the rabbit. *Am J Physiol.* 1967;213(4):1031–1038.

18. Sonnenberg H, Solomon S, Frazier DT. Sodium and chloride movement into the central canal of cat spinal cord. *Proc Soc Exp Biol Med.* 1967;124(4): 1316–1320.

19. Bradbury MW. Physiopathology of the blood-brain barrier. *Adv Exp Med Biol.* 1976;69:507–516.

20. Tait MJ, Saadoun S, Bell BA, Papadopoulos MC. Water movements in the brain: role of aquaporins. *Trends Neurosci.* 2008;31(1):37–43.

21. MacAulay N, Zeuthen T. Water transport between CNS compartments: contributions of aquaporins and cotransporters. *Neuroscience.* 2010;168(4): 941–956.

22. Papadopoulos MC, Verkman AS. Aquaporin water channels in the nervous system. *Nat Rev Neurosci.* 2013;14(4):265–277.

23. Cserr HF. Role of secretion and bulk flow of brain interstitial fluid in brain volume regulation. *Ann NY Acad Sci.* 1988;529:9–20.

24. Bering Jr EA. Water exchange of central nervous system and cerebrospinal fluid. *J Neurosurg.* 1952;9(3):275–287.

25. Bateman GA. Extending the hydrodynamic hypothesis in chronic hydrocephalus. *Neurosurg Rev.* 2005;28(4):333–334.

26. Bateman GA, Brown KM. The measurement of CSF flow through the aqueduct in normal and hydrocephalic children: from where does it come, to where does it go? *Childs Nerv Syst.* 2012;28(1):55–63.

27. Johanson CE, Duncan III JA, Klinge PM, Brinker T, Stopa EG, Silverberg GD. Multiplicity of cerebrospinal fluid functions: new challenges in health and disease. *Cerebrospinal Fluid Res.* 2008;5:10.

28. Masserman JH. Cerebrospinal hydrodynamics. IV. Clinical experimental studies. *Arch Neurol Psychiatry.* 1934;32:523–553.

29. Heisey SR, Held D, Pappenheimer JR. Bulk flow and diffusion in the cerebrospinal fluid system of the goat. *Am J Physiol.* 1962;203:775–781.

30. Fishman RA. The cerebrospinal fluid production rate is reduced in dementia of the Alzheimer's type. *Neurology.* 2002;58(12):1866.

31. Ekstedt J. CSF hydrodynamic studies in man. 2. Normal hydrodynamic variables related to CSF pressure and flow. *J Neurol Neurosurg Psychiatry.* 1978;41(4):345–353.

32. Rubin RC, Henderson ES, Ommaya AK, Walker MD, Rall DP. The production of cerebrospinal fluid in man and its modification by acetazolamide. *J Neurosurg.* 1966;25(4):430–436.

33. Cutler RW, Page L, Galicich J, Watters GV. Formation and absorption of cerebrospinal fluid in man. *Brain*. 1968;91(4):707–720.
34. Rottenberg DA, Howieson J, Deck MD. The rate of CSF formation in man: preliminary observations on metrizamide washout as a measure of CSF bulk flow. *Ann Neurol*. 1977;2(6):503–510.
35. Rottenberg DA, Deck MD, Allen JC. Metrizamide washout as a measure of CSF bulk flow. *Neuroradiology*. 1978;16:203–206.
36. Oreskovic D, Marakovic J, Vukic M, Rados M, Klarica M. Fluid perfusion as a method of cerebrospinal fluid formation rate – critical appraisal. *Coll Antropol*. 2008;32(suppl 1):133–137.
37. Oreskovic D, Klarica M. The formation of cerebrospinal fluid: nearly a hundred years of interpretations and misinterpretations. *Brain Res Rev*. 2010;64(2): 241–262.
38. Bulat M, Lupret V, Oreskovic D, Klarica M. Transventricular and transpial absorption of cerebrospinal fluid into cerebral microvessels. *Coll Antropol*. 2008;32(suppl 1):43–50.
39. Bulat M, Klarica M. Recent insights into a new hydrodynamics of the cerebrospinal fluid. *Brain Res Rev*. 2011;65(2):99–112.
40. O'Donnell M. NMR blood flow imaging using multiecho, phase contrast sequences. *Med Phys*. 1985;12(1):59–64.
41. Bradley Jr WG, Kortman KE, Burgoyne B. Flowing cerebrospinal fluid in normal and hydrocephalic states: appearance on MR images. *Radiology*. 1986;159(3):611–616.
42. Feinberg DA, Mark AS. Human brain motion and cerebrospinal fluid circulation demonstrated with MR velocity imaging. *Radiology*. 1987;163(3): 793–799.
43. Nitz WR, Bradley Jr WG, Watanabe AS, et al. Flow dynamics of cerebrospinal fluid: assessment with phase-contrast velocity MR imaging performed with retrospective cardiac gating. *Radiology*. 1992;183(2):395–405.
44. Bradley Jr WG, Scalzo D, Queralt J, Nitz WN, Atkinson DJ, Wong P. Normal-pressure hydrocephalus: evaluation with cerebrospinal fluid flow measurements at MR imaging. *Radiology*. 1996;198(2):523–529.
45. Gideon P, Thomsen C, Stahlberg F, Henriksen O. Cerebrospinal fluid production and dynamics in normal aging: a MRI phase-mapping study. *Acta Neurol Scand*. 1994;89(5):362–366.
46. Huang TY, Chung HW, Chen MY, et al. Supratentorial cerebrospinal fluid production rate in healthy adults: quantification with two-dimensional cine phase-contrast MR imaging with high temporal and spatial resolution. *Radiology*. 2004;233(2):603–608.
47. Piechnik SK, Summers PE, Jezzard P, Byrne JV. Magnetic resonance measurement of blood and CSF flow rates with phase contrast – normal values, repeatability and CO_2 reactivity. *Acta Neurochir Suppl*. 2008;102:263–270.
48. Yoshida K, Takahashi H, Saijo M, et al. Phase-contrast MR studies of CSF flow rate in the cerebral aqueduct and cervical subarachnoid space with correlation-based segmentation. *Magn Reson Med Sci*. 2009;8(3):91–100.
49. Penn RD, Basati S, Sweetman B, Guo X, Linninger A. Ventricle wall movements and cerebrospinal fluid flow in hydrocephalus. *J Neurosurg*. 2011;115(1): 159–164.

50. Stone SS, Warf BC. Combined endoscopic third ventriculostomy and choroid plexus cauterization as primary treatment for infant hydrocephalus: a prospective North American series. *J Neurosurg Pediatr.* 2014;14(5):439–446.

51. Warf BC. Congenital idiopathic hydrocephalus of infancy: the results of treatment by endoscopic third ventriculostomy with or without choroid plexus cauterization and suggestions for how it works. *Childs Nerv Syst.* 2013;29(6):935–940.

52. Szmydynger-Chodobska J, Strazielle N, Gandy JR, et al. Posttraumatic invasion of monocytes across the blood-cerebrospinal fluid barrier. *J Cereb Blood Flow Metab.* 2012;32(1):93–104.

53. Levy YS, Gilgun-Sherki Y, Melamed E, Offen D. Therapeutic potential of neurotrophic factors in neurodegenerative diseases. *Bio Drugs.* 2005;19(2):97–127.

54. Fumagalli F, Molteni R, Calabrese F, Maj PF, Racagni G, Riva MA. Neurotrophic factors in neurodegenerative disorders : potential for therapy. *CNS Drugs.* 2008;22(12):1005–1019.

55. Ruozi B, Belletti D, Bondioli L, et al. Neurotrophic factors and neurodegenerative diseases: a delivery issue. *Int Rev Neurobiol.* 2012;102:207–247.

56. Lanni C, Stanga S, Racchi M, Govoni S. The expanding universe of neurotrophic factors: therapeutic potential in aging and age-associated disorders. *Curr Pharm Des.* 2010;16(6):698–717.

57. Gaddam SK, Cruz J, Robertson C. Erythropoietin and cytoprotective cytokines in experimental traumatic brain injury. *Methods Mol Biol.* 2013;982: 141–162.

58. Bornstein N, Poon WS. Accelerated recovery from acute brain injuries: clinical efficacy of neurotrophic treatment in stroke and traumatic brain injuries. *Drugs Today Barc.* 2012;48(suppl A):43–61.

59. Emerich DF, Borlongan CV. Potential of choroid plexus epithelial cell grafts for neuroprotection in Huntington's disease: what remains before considering clinical trials. *Neurotoxicity Res.* 2009;15(3):205–211.

60. Redzic ZB, Preston JE, Duncan JA, Chodobski A, Szmydynger-Chodobska J. The choroid plexus-cerebrospinal fluid system: from development to aging. *Curr Top Dev Biol.* 2005;71:1–52.

61. Spector R, Johanson CE. The nexus of vitamin homeostasis and DNA synthesis and modification in mammalian brain. *Mol Brain.* 2014;7:3.

62. Johanson C, Stopa E, McMillan P, Roth D, Funk J, Krinke G. The distributional nexus of choroid plexus to cerebrospinal fluid, ependyma and brain: toxicologic/pathologic phenomena, periventricular destabilization, and lesion spread. *Toxicol Pathol.* 2011;39(1):186–212.

63. Chodobski A, Szmydynger-Chodobska J. Choroid plexus: target for polypeptides and site of their synthesis. *Microsc Res Tech.* 2001;52(1):65–82.

64. Lo EH, Singhal AB, Torchilin VP, Abbott NJ. Drug delivery to damaged brain. *Brain Res Rev.* 2001;38(1–2):140–148.

65. Begley DJ. Delivery of therapeutic agents to the central nervous system: the problems and the possibilities. *Pharmacol Therap.* 2004;104(1):29–45.

66. Pardridge WM. Drug transport in brain via the cerebrospinal fluid. *Fluids Barriers CNS.* 2011;8(1):7.

67. Pardridge WM. Drug transport across the blood-brain barrier. *J Cereb Blood Flow Metab.* 2012;32(11):1959–1972.

68. Johanson CE, Stopa EG, McMillan PN, The Blood–Cerebrospinal Fluid Barrier: Structure and Functional Significance. In: Nag S, ed. *The Blood–Brain and Other Neural Barriers: Reviews and Protocols, Methods in Molecular Biology*, 686. Springer; 2010:101–131.

69. Blasberg RG, Patlak C, Fenstermacher JD. Intrathecal chemotherapy: brain tissue profiles after ventriculocisternal perfusion. *J Pharmacol Exp Ther.* 1975;195(1):73–83.

70. Thorne RG, Nicholson C. *In vivo* diffusion analysis with quantum dots and dextrans predicts the width of brain extracellular space. *Proc Natl Acad Sci USA.* 2006;103(14):5567–5572.

71. Thorne RG, Frey WH. Delivery of neurotrophic factors to the central nervous system: pharmacokinetic considerations. *Clin Pharmacokinet.* 2001;40(12): 907–946.

72. Krewson CE, Klarman ML, Saltzman WM. Distribution of nerve growth factor following direct delivery to brain interstitium. *Brain Res.* 1995;680(1–2): 196–206.

73. Saltzman WM, Mak MW, Mahoney MJ, Duenas ET, Cleland JL. Intracranial delivery of recombinant nerve growth factor: release kinetics and protein distribution for three delivery systems. *Pharm Res.* 1999;16(2): 232–240.

74. Aird RB. A study of intrathecal, cerebrospinal fluid-to-brain exchange. *Exp Neurol.* 1984;86(2):342–358.

75. Pardridge WM. *Peptide drug delivery to the brain.* New York: Raven Press; 1991.

76. Venero JL, Hefti F, Knusel B. Trophic effect of exogenous nerve growth factor on rat striatal cholinergic neurons: comparison between intraparenchymal and intraventricular administration. *Mol Pharmacol.* 1996;49(2): 303–310.

77. Proescholdt MG, Hutto B, Brady LS, Herkenham M. Studies of cerebrospinal fluid flow and penetration into brain following lateral ventricle and cisterna magna injections of the tracer [14C]inulin in rat. *Neuroscience.* 2000;95(2): 577–592.

78. Emmett CJ, Stewart GR, Johnson RM, Aswani SP, Chan RL, Jakeman LB. Distribution of radioiodinated recombinant human nerve growth factor in primate brain following intracerebroventricular infusion. *Exp Neurol.* 1996;140(2): 151–160.

79. Bielicki J, McIntyre C, Anson DS. Comparison of ventricular and intravenous lentiviral-mediated gene therapy for murine MPS VII. *Mol Genet Metab.* 2010;101(4):370–382.

80. Regev L, Ezrielev E, Gershon E, Gil S, Chen A. Genetic approach for intracerebroventricular delivery. *Proc Natl Acad Sci USA.* 2010;107(9): 4424–4429.

81. Fenstermacher JD, Ghersi-Egea JF, Finnegan W, Chen JL. The rapid flow of cerebrospinal fluid from ventricles to cisterns via subarachnoid velae in the normal rat. *Acta Neurochir Suppl.* 1997;70:285–287.

82. Weed LH. Meninges and cerebrospinal fluid. *J Anat.* 1938;72(Pt 2):181–215.

83. Brinker T, Stopa E, Morrison J, Klinge P. A new look at cerebrospinal fluid circulation. *Fluids Barriers CNS.* 2014;11:10.

84. Rennels ML, Gregory TF, Blaumanis OR, Fujimoto K, Grady PA. Evidence for a 'paravascular' fluid circulation in the mammalian central nervous system, provided by the rapid distribution of tracer protein throughout the brain from the subarachnoid space. *Brain Res.* 1985;326(1):47–63.

85. Guan J, Beilharz EJ, Skinner SJ, Williams CE, Gluckman PD. Intracerebral transportation and cellular localisation of insulin-like growth factor-1 following central administration to rats with hypoxic-ischemic brain injury. *Brain Res.* 2000;853(2):163–173.

86. Iliff JJ, Wang M, Liao Y, et al. A paravascular pathway facilitates CSF flow through the brain parenchyma and the clearance of interstitial solutes, including amyloid beta. *Sci Transl Med.* 2012;4(147):147.

87. Abbott NJ. Evidence for bulk flow of brain interstitial fluid: significance for physiology and pathology. *Neurochem Int.* 2004;45(4):545–552.

88. Hadaczek P, Yamashita Y, Mirek H, et al. The "perivascular pump" driven by arterial pulsation is a powerful mechanism for the distribution of therapeutic molecules within the brain. *Mol Ther.* 2006;14(1):69–78.

89. Veening JG, Barendregt HP. The regulation of brain states by neuroactive substances distributed via the cerebrospinal fluid; a review. *Cerebrospinal Fluid Res.* 2010;7(1):1.

90. Zhang ET, Inman CB, Weller RO. Interrelationships of the pia mater and the perivascular (Virchow-Robin) spaces in the human cerebrum. *J Anat.* 1990;170:111–123.

91. Krisch B, Leonhardt H, Oksche A. Compartments and perivascular arrangement of the meninges covering the cerebral cortex of the rat. *Cell Tissue Res.* 1984;238(3):459–474.

92. Bechmann I, Galea I, Perry VH. What is the blood-brain barrier (not)? *Trends Immunol.* 2007;28(1):5–11.

93. Krueger M, Bechmann I. CNS pericytes: concepts, misconceptions, and a way out. *Glia.* 2010;58(1):1–10.

94. Hurford R, Charidimou A, Fox Z, Cipolotti L, Jager R, Werring DJ. MRI-visible perivascular spaces: relationship to cognition and small vessel disease MRI markers in ischaemic stroke and TIA. *J Neurol Neurosurg Psychiatry.* 2014;85(5):522–525.

95. Kwee RM, Kwee TC. Virchow-Robin spaces at MR imaging. *Radiographics.* 2007;27(4):1071–1086.

96. Borlongan CV, Skinner SJM, Geaney M, Vasconcellos AV, Elliott RB, Emerich DF. Intracerebral transplantation of porcine choroid plexus provides structural and functional neuroprotection in a rodent model of stroke. *Stroke.* 2004;35(9):2206–2210.

97. Borlongan CV, Skinner SJ, Geaney M, Vasconcellos AV, Elliott RB, Emerich DF. Intracerebral transplantation of porcine choroid plexus provides structural and functional neuroprotection in a rodent model of stroke. *Stroke.* 2004;35(9):2206–2210.

98. Tseng JL, Aebischer P. Encapsulated neural transplants. *Prog Brain Res.* 2000;127:189–202.

99. Murua A, Portero A, Orive G, Hernández RM, de Castro M, Pedraz JL. Cell microencapsulation technology: towards clinical application. *J Control Release.* 2008;132(2):76–83.

100. Lindvall O, Wahlberg LU. Encapsulated cell biodelivery of GDNF: a novel clinical strategy for neuroprotection and neuroregeneration in Parkinson's disease? *Exp Neurol.* 2008;209(1):82–88.

101. Heile AMB, Wallrapp C, Klinge PM, et al. Cerebral transplantation of encapsulated mesenchymal stem cells improves cellular pathology after experimental traumatic brain injury. *Neurosci Lett.* 2009;463(3):176–181.

102. Glage S, Klinge PM, Miller MC, et al. Therapeutic concentrations of glucagon-like peptide-1 in cerebrospinal fluid following cell-based delivery into the cerebral ventricles of cats. *Fluids Barriers CNS.* 2011;8:18.

103. Thanos CG, Bintz BE, Emerich DF. Stability of alginate-polyornithine microcapsules is profoundly dependent on the site of transplantation. *J Biomed Mater Res A.* 2007;81(1):1–11.

104. Winn SR, Lindner MD, Lee A, Haggett G, Francis JM, Emerich DF. Polymer-encapsulated genetically modified cells continue to secrete human nerve growth factor for over one year in rat ventricles: behavioral and anatomical consequences. *Exp Neurol.* 1996;140(2):126–138.

105. Wahlberg LU, Lind G, Almqvist PM, et al. Targeted delivery of nerve growth factor via encapsulated cell biodelivery in Alzheimer disease: a technology platform for restorative neurosurgery. *J Neurosurg.* 2012;117(2):340–347.

106. Borlongan CV, Skinner SJ, Geaney M, Vasconcellos AV, Elliott RB, Emerich DF. Neuroprotection by encapsulated choroid plexus in a rodent model of Huntington's disease. *Neuroreport.* 2004;15(16):2521–2525.

107. Redzic ZB. Studies on the human choroid plexus *in vitro. Fluids Barriers CNS.* 2013;10(1):10.

108. Nauta AJ, Fibbe WE. Immunomodulatory properties of mesenchymal stromal cells. *Blood.* 2007;110(10):3499–3506.

109. Ohtaki H, Ylostalo JH, Foraker JE, et al. Stem/progenitor cells from bone marrow decrease neuronal death in global ischemia by modulation of inflammatory/immune responses. *Proc Natl Acad Sci USA.* 2008;105(38):14638–14643.

110. Kassis I, Vaknin-Dembinsky A, Karussis D. Bone marrow mesenchymal stem cells: agents of immunomodulation and neuroprotection. *Curr Stem Cell Res Ther.* 2011;6(1):63–68.

111. Meyerrose T, Olson S, Pontow S, et al. Mesenchymal stem cells for the sustained *in vivo* delivery of bioactive factors. *Adv Drug Deliv Rev.* 2010;62(12):1167–1174.

112. Sanberg PR, Eve DJ, Cruz LE, Borlongan CV. Neurological disorders and the potential role for stem cells as a therapy. *Br Med Bull.* 2012;101:163–181.

113. Caplan AI, Correa D. The MSC: an injury drugstore. *Cell Stem Cell.* 2011;9(1):11–15.

114. Borlongan CV. Cell therapy for stroke: remaining issues to address before embarking on clinical trials. *Stroke.* 2009;40(3 suppl):S146–S148.

115. Zhang W, He X. Microencapsulating and banking living cells for cell-based medicine. *J Healthc Eng.* 2011;2(4):427–446.

116. Greco SJ, Rameshwar P. Mesenchymal stem cells in drug/gene delivery: implications for cell therapy. *Ther Deliv.* 2012;3(8):997–1004.

117. van Velthoven CT, Kavelaars A, van Bel F, Heijnen CJ. Regeneration of the ischemic brain by engineered stem cells: fuelling endogenous repair processes. *Brain Res Rev.* 2009;61(1):1–13.

118. Bliss T, Guzman R, Daadi M, Steinberg GK. Cell transplantation therapy for stroke. *Stroke.* 2007;38(2 suppl):817–826.
119. Mejia-Toiber J, Castillo CG, Giordano M. Strategies for the development of cell lines for *ex vivo* gene therapy in the central nervous system. *Cell Transplant.* 2011;20(7):983–1001.
120. Joyce N, Annett G, Wirthlin L, Olson S, Bauer G, Nolta JA. Mesenchymal stem cells for the treatment of neurodegenerative disease. *Regen Med.* 2010;5(6):933–946.
121. Sadan O, Shemesh N, Barzilay R, et al. Mesenchymal stem cells induced to secrete neurotrophic factors attenuate quinolinic acid toxicity: a potential therapy for Huntington's disease. *Exp Neurol.* 2012;234(2):417–427.
122. Klinge PM, Harmening K, Miller MC, et al. Encapsulated native and glucagon-like peptide-1 transfected human mesenchymal stem cells in a transgenic mouse model of Alzheimer's disease. *Neurosci Lett.* 2011;497(1):6–10.
123. Knippenberg S, Thau N, Dengler R, Brinker T, Petri S. Intracerebroventricular injection of encapsulated human mesenchymal cells producing glucagon-like peptide-1 prolongs survival in a mouse model of ALS. *PLoS One.* 2012;7(6):e36857.
124. Heile A, Brinker T. Clinical translation of stem cell therapy in traumatic brain injury: the potential of encapsulated mesenchymal cell biodelivery of glucagon-like peptide-1. *Dialogues Clin Neurosci.* 2011;13(3):279–286.
125. Eriksdotter-Jonhagen M, Linderoth B, Lind G, et al. Encapsulated cell biodelivery of nerve growth factor to the Basal forebrain in patients with Alzheimer's disease. *Dement Geriatr Cogn Disord.* 2012;33(1):18–28.
126. Emborg ME, Kordower JH. Delivery of therapeutic molecules into the CNS. *Prog Brain Res.* 2000;128:323–332.
127. Winkler J, Ramirez GA, Thal LJ, Waite JJ. Nerve growth factor (NGF) augments cortical and hippocampal cholinergic functioning after p75NGF receptor-mediated deafferentation but impairs inhibitory avoidance and induces fear-related behaviors. *J Neurosci.* 2000;20(2):834–844.
128. Day-Lollini PA, Stewart GR, Taylor MJ, Johnson RM, Chellman GJ. Hyperplastic changes within the leptomeninges of the rat and monkey in response to chronic intracerebroventricular infusion of nerve growth factor. *Exp Neurol.* 1997;145(1):24–37.
129. Yamada K, Kinoshita A, Kohmura E, et al. Basic fibroblast growth factor prevents thalamic degeneration after cortical infarction. *J Cereb Blood Flow Metab.* 1991;11(3):472–478.
130. Johanson CE, Szmydynger-Chodobska J, Chodobski A, Baird A, McMillan P, Stopa EG. Altered formation and bulk absorption of cerebrospinal fluid in FGF-2-induced hydrocephalus. *Am J Physiol.* 1999;277(1 Pt 2):263–271.
131. Pearce RK, Collins P, Jenner P, Emmett C, Marsden CD. Intraventricular infusion of basic fibroblast growth factor (bFGF) in the MPTP-treated common marmoset. *Synapse.* 1996;23(3):192–200.

Choroid Plexus Tumors

Sean A. Grimm*,
Marc C. Chamberlain†

*Cadence Health Brain and Spine Tumor Center, Warrenville,
IL, USA; †Department of Neurology and Neurological Surgery,
Seattle Cancer Care Alliance, Fred Hutchinson Cancer Research
Center, University of Washington, Seattle, WA, USA

INTRODUCTION

The choroid plexus (CP) consists of a cuboidal cell monolayer of epithelium, blood vessels, and interstitial connective tissue.[1] The choroidal epithelium is composed of a single row of cuboidal

The Choroid Plexus and Cerebrospinal Fluid. http://dx.doi.org/10.1016/B978-0-12-801740-1.00005-6

epithelial cells, arranged into villi around a core of blood vessels and connective tissue.[1] Topographically, the CP is located on the floor of the lateral, third, and fourth ventricles and can occasionally develop in atopic locations. It produces the majority of the cerebrospinal fluid (CSF). Though rare, the CP can be the site of primary or secondary tumors.

PRIMARY CHOROID PLEXUS TUMORS

The World Health Organization (WHO) classification of central nervous system (CNS) tumors recognizes three primary tumors that arise from the CP: choroid plexus papilloma (CPP, WHO grade I), atypical choroid plexus papilloma (ACPP, WHO grade II), and choroid plexus carcinoma (CPC, WHO grade III) (Table 5.1).

The low grade papillomas outnumber carcinomas by a ratio of at least 5:1.[2] ACPPs are least common and their clinical correlates have not been described, so their discussion will be necessarily limited. In the vast majority of cases, CPPs are benign and slow growing tumors. In contrast, CPCs are highly malignant tumors with dismal prognosis. CPCs are frequently found in families with Lie–Fraumeni syndrome (a familial syndrome characterized by germline mutations in the *TP53* suppressor gene resulting in a heterogeneous phenotype of early-onset cancers).

CPTs predominantly occur in children, although adult cases are rarely encountered. They represent 0.3–0.6% of all brain tumors, 2–4% of those that occur in patients under 15 years of age, 10–20% of those that present in the first year of a patient's life.[2] Rarely, they may be diagnosed by ultrasound *in utero*. In children, these tumors most commonly arise in the lateral ventricle, followed by the fourth ventricle, and less commonly, the third ventricle.[3–6] In rare instances these tumors arise from the cerebellopontine angle, or from ectopic CP tissue in the brain parenchyma, sacral canal, pineal region, or cerebellar hemispheres.[3,7–12] There is a striking age dependency with respect to tumor topography, as in adults the majority of CPTs are located in the fourth ventricle. Median age at presentation is 1.5, 1.5, 22.5, and 35.5 years in tumors located in the lateral ventricles, third ventricle, fourth ventricle, and cerebellarpontine angle, respectively.[13]

Due to of their origin in noneloquent brain, most primary CPTs reach considerable size before radiologic diagnosis (median 5.0 cm in diameter in one series).[14] The most common presenting

TABLE 5.1   Primary Choroid Plexus Tumors

General features	Choroid plexus papilloma (WHO grade I)	Atypical choroid plexus papilloma (WHO grade II)	Choroid plexus carcinoma (WHO grade III)
General histological features	Fibrovascular stalk surrounded by single layer of cuboidal to epithelium arranged in papillary configuration	Intermediate histology; characterized by increased mitotic activity compared with CPP	At least four of the following: increased cellularity, blurring of papillary architecture, high mitotic activity, nuclear pleomorphism, and necrosis
Mitotic activity	<2/10 HPF	>2/10 HPF	>5/10 HPF
Radiographic features	Well circumscribed intraventricular mass with cauliflower appearance and contrast enhancement; iso- or hyperdense on CT and iso- or hypointense on T1-weighted MRI Calcification in 25%	Similar to CPP but may possess irregular or invasive margins with associated edema	Generally larger than CPPs and frequently invade adjacent parenchyma with associated edema; heterogeneous contrast enhancement with possible calcifications, hemorrhage, necrosis, or leptomeningeal enhancement (dissemination)

WHO, World Health Organization; HPF, high power field; CPP, choroid plexus papilloma; CT, computerized tomography; MRI, magnetic resonance imaging.

symptoms are headache and nausea/vomiting caused by increased intracranial pressure from obstructive hydrocephalus. Patients with CPCs frequently have leptomeningeal metastases at diagnosis (up to 20% in some series) and even the benign CPP can be present with CSF dissemination (<5%).[15,16] Consequently, all patients with a CPT should undergo magnetic resonance imaging (MRI) with gadolinium contrast of their entire craniospinal axis as part of the initial staging workup. In addition, a diagnostic lumbar puncture for CSF cytology is recommended.

Radiologic Appearance

CPP and CPC may have similar appearance on neuroimaging. On brain computerized tomography (CT), they appear isodense to slightly hypodense, and may be heavily calcified (particularly CPPs). On MRI, CPTs are T1-hypointense or iso-intense and T2-hyperintense; flow voids are common.[17,18] Both tumors display homogenous enhancement following the administration of gadolinium contrast. A heterogeneous lesion with necrosis, parenchymal invasion, and edema is more suggestive of a CPC.[18] A well-demarcated lesion with a thin iso-intense (transparent) capsule surrounding the tumor on T1- and T2-weighted images and absence of cystic necrosis is typical of CPP.[17] CPP and CPC exhibit hyper-perfusion on DSC perfusion with relative cerebral blood volumes greater than three.[18]

The differential diagnostic considerations for an intraventricular mass on imaging includes CPP, CPC, metastasis, ependymoma, subependymoma, meningioma, central neurocytoma, pilocytic astrocytoma, villous hypertrophy of the chorid plexus, embryonal (primitive neuroectodermal) tumors, germ cell tumors, and solitary fibrous tumors. Histiocytosis, infectious choroid plexitis (cryptococcus neoformans, tuberculosis, nocardiosis, cytomegalovirus, and toxoplasmosis), noninfectious choroid plexitis (Wegner's granulomatosis), and CP enlargement in Sturge–Weber syndrome are additional diagnostic considerations.[18-26] Because the radiologic features are nonspecific, surgical resection, or biopsy for histology is essential to make a diagnosis of a primary CPT. Special attention to histology needs to be made in differentiating CPTs from ependymoma (papillary variant), meningioma (papillary variant), villous hypertrophy, and atypical teratoid/rhabdoid tumors (AT/RT).[27]

Pathology

Macroscopic Appearance

On gross inspection, CPP typically appears as an intraventricular, globular, firm, reddish-brown, or pinkish mass that appears as a "cauliflower floating in the cerebrospinal fluid (CSF)." The tumor is usually well demarcated from normal brain tissue. In contrast,

CPC frequently presents itself as a large mass filling one or both ventricles, and is associated with parenchymal invasion.[28]

Microscopic Appearance

CPP is classified as a WHO grade I tumor. In majority of the cases, CPP faithfully replicates the architecture of nonneoplastic CP, differing only in the magnitude of growth.[28] Delicate fibrovascular connective tissue fronds are covered by a single layer of uniform cuboidal to columnar epithelial cells with round or oval monomorphic nuclei.[2] The mitotic activity is low and Ki-67/MIB-1 labeling has been reported as 1.9% (range, 0.2–6%). Brain invasion, high cellularity, necrosis, and nuclear pleomorphism are absent. The presence of mitotic figures on the initial histology may predict a higher likelihood or recurrence or malignant evolution.[29] In extremely rare cases, there may be bone formation or neuromelanin production in CPPs.[6,30] The main differential considerations are normal CP and villous hypertrophy. Villous hypertrophy is a benign enlargement of normal CP tissue in the lateral ventricles that has normal CP histology and is distinguished from primary CPTs by utilizing proliferation markers.[27,31–33] Recent findings have implicated Notch3 signaling, the transcription factor TWIST1, platelet-derived growth factor receptor, and the tumor necrosis factor-related apoptosis-inducing ligand pathway in CPP tumorgenesis.[34]

CPC is a malignant tumor classified as a WHO grade III tumor. CPC can be difficult to diagnose through histology. The tumor shows obvious features of malignancy, which may include the following features: frequent mitoses (greater than 5/10 HPF), nuclear pleomorphism, blurring of the papillary pattern with poorly structured sheets of tumor cells, and necrotic areas.[2] Parenchymal brain invasion is common. On immunohistochemical staining, CPCs express cytokeratins and about 20% are positive for glial fibrillary acidic protein. Epithelial membrane antigen (EMA) is usually not expressed. Mean Ki67/MIB-1 labeling index is 13.8–18.5%.[2] The main differential considerations include papillary variant ependymoma and metastasis form system cancer in adults, and papillary variant ependymoma, papillary variant meningioma, germ cell tumors, and embryonal tumors (particularly AT/RT) in the pediatric population. Because of the absence of a highly specific marker for CPTs, common occurrence of metastatic adenocarcinoma from an unknown primary, and the rarity of primary CPTs, a definitive prospective diagnosis of CPC is challenging in adults.[28] The papillary variant of ependymoma (a high grade, WHO grade III tumor) can be differentiated from CPTs by typical ependymoma features, which include ependymal rosettes and perivascular pseudorosettes[2]). Papillary meningioma (a high grade, WHO grade III tumor) tends to occur in pediatric patients and displays a pseudo-papillary pattern in the majority of the tumor.[2] Vimentin positivity is found in all meningiomas and EMA may be present. AT/RT is a malignant tumor found in the pediatric population that usually has groups of rhabdoid cells with eccentric, pleomorphic nuclei, and abundant eosinophillic cytoplasm. Most tumors contain variable components with primitive neuroectodermal, mesenchymal (less common), and epithelial differentiation (least common). The epithelial differentiation takes the form of papillary structures making the distinction from CPC challenging.[2]

Treatment

Surgery

Maximal safe surgical resection is the primary treatment modality for CPTs.[35,36] Complete resection of the primary intraventricular tumor can be achieved in most cases.[3,5,37–39] Patients with obstructive hydrocephalus may require permanent CSF diversion by placement of a ventriculoperitoneal shunt or endoscopic third ventriculostomy.

Radiotherapy/Chemotherapy

CPP

Adjuvant radiotherapy is not indicated, even in instances of subtotal resection.[13,40] This is particularly true in young children, where radiotherapy has a detrimental effect on the developing brain. In those with tumor recurrence, repeat resection, or adjuvant radiation therapy (RT) (50–54 Gy in 20–25 fractions) is recommended. There is no role for chemotherapy in patients with CPP at diagnosis or recurrence.

CPC

In patients with CPC and an incomplete resection, a second surgical resection should be considered, as often advocated for ependymomas.[41,42] Improved survival is seen in patients having undergone image verified complete resection as compared to a partial resection.[36] Because of the high risk of recurrence even after complete resection, adjuvant RT or chemotherapy is recommended. Prospective studies comparing or demonstrating benefit of these adjuvant treatment modalities are lacking. Adjuvant RT is associated with improved survival in retrospective studies.[13] Patients with CSF dissemination should receive craniospinal axis radiotherapy (CSI), while the type of RT field (CSI vs. involved field) in those without metastatic disease is controversial.[37,43] Given the detrimental effects of RT to the developing brain and the risk of late neurologic sequelae in patients <3 years of age, RT sparing treatment regimens should be considered.[44] Patients with wild type *TP53* tumor suppressor gene may be offered an RT sparing treatment regimen (see the discussion regarding *TP53* in "Prognosis").[45]

A meta-analysis of 857 documented cases of CPT suggested a survival benefit in patients with CPC treated with adjuvant chemotherapy.[46] Eight of 22 patients had a radiographic response to adjuvant chemotherapy in another meta-analysis.[13] Numerous chemotherapeutic agents have been advocated, although prospective studies comparing agents are lacking. One retrospective study suggested that there was a survival benefit for treatment with cyclophosphamide, etoposide, and carboplatin, while the effect of vincristine was found to be marginally significant.[47] Caution must be exercised when extrapolating pediatric schedules to an adult patient given the higher susceptibility of adults to myelosuppression.[48]

Prognosis

CPP

Pediatric patients with CPP have an excellent prognosis. Complete surgical excision can be curative.[3,37,40,49] A single institution series of 41 patients with CPP reported 5-year local control, distant brain control, and overall survival of 84, 92, and 97%, respectively.[24] Comparison of gross total resection (GTR) and subtotal resection (STR) at 5 years showed a significant increase in local control (100% vs. 68%). A meta-analysis of 566 CPTs reported 1-, 5-, and 10-year projected survival rates of 90, 81, and 77% for patients with CPPs.[13] Comparison of gross total resection and biopsy at 10 years showed an increase in survival (85% vs. 56%).

A metaanalysis of 193 adult patients (median age 40 years) reported a significant increase in progression-free survival (median

11 years vs. 5 years) and overall survival (mean 21.5 years vs. 9.8 years) in those who had GTR compared to STR.[35] The 1-, 5-, and 10-year control rates were 100, 72, and 41% following GTR, compared to 85, 32, and 22% following STR, respectively.

Life long surveillance in adults and children is recommended, even after GTR. Those with mitotic figures on initial pathology may have higher likelihood or recurrence or malignant evolution, should be followed more closely.[29,35]

CPC

The 5-year local control, distant brain control, and overall survival was 71, 41, and 35% in the meta-analysis referenced earlier.[13] GTR and adjuvant RT was associated with improved survival. Patients with GTR had a 2-year survival rate of 72% compared to 34% with STR.

Mutation of the TP53 tumor suppressor gene has been proposed as a prognostic marker for CPCs.[45] Five-year overall survival rates for patients with mutated and wild-type TP53 were 0 and 82%, respectively. Fourteen of 16 CPC patients with wild-type TP53 were alive at 5 years without having received adjuvant radiotherapy. The authors of this study have proposed a new approach to the management of children with CPC. At surgery, comprehensive family and clinical histories should be taken. If this information suggests the possibility of Lie–Fraumeni syndrome, then TP53 mutation analysis should be performed. Patients with wild-type tumors should be offered an RT-sparing protocol. Those with mutated TP53 should be considered for novel therapies.[45]

SECONDARY CHOROID PLEXUS TUMORS

Any systemic malignancy has the propensity to metastasize to the CP. One prospective autopsy study found metastases to this region in 2.6% of patients with cancer.[50] A retrospective study suggests that renal cell carcinoma may have a particular propensity to metastasize to the CP.[51] CP metastases are single or multiple and usually asymptomatic but may cause hydrocephalus with resultant increase in intracranial pressure by causing obstruction of outflow at the foramen of Monro. CP metastases often disseminate in the CSF resulting in leptomeningeal metastases. The finding of an avidly enhancing mass arising from the CP makes the diagnosis on brain imaging. Early on, the imaging changes may be subtle with only asymmetry of the contrast enhancing choroid in the lateral ventricles. Often, the diagnosis is only made after noting a progressive enlargement in CP enhancement on serial imaging. Without a high index of suspicion, the metastasis may be mistaken for a benign tumor such as an intraventricular meningioma or prominent normal CP.[52] Because of the high risk of CSF dissemination, all patients with CP metastases should undergo whole spine imaging with and without gadolinium contrast and CSF sampling to assess for malignant cells. The treatment of CP metastases is similar to other brain metastases and includes stereotactic radiosurgery, whole brain radiotherapy, and systemic chemotherapy. The choice of a particular modality depends on the patient's clinical status and the primary cancer type.

References

1. Ra F. *Cerebrospinal Fluid in Diseases of the Nervous System.* 2nd ed. Philadelphia, PA: W. B Saunders Company; 1992.
2. Louis DN, Phgaki H, Wiester OD, Webster KC, Burger PC, Jouvet A, Scheithauer BW, Kleihues P. The 2007 WHO Classification of Tumors of the Central Nervous System. *Acta Neuropathol.* 2007;114(2):97–109.
3. Bettegowda C, Adogwa O, Mehta V, et al. Treatment of choroid plexus tumors: a 20-year single institutional experience. *J Neurosurg Pediatr.* 2012;10(5):398–405.
4. Due-Tonnessen B, Helseth E, Skullerud K, Lundar T. Choroid plexus tumors in children and young adults: report of 16 consecutive cases. *Childs Nerv Syst.* 2001;17(4–5):252–256.
5. Jaiswal AK, Jaiswal S, Sahu RN, Das KB, Jain VK, Behari S. Choroid plexus papilloma in children: diagnostic and surgical considerations. *J Pediatr Neurosci.* 2009;4(1):10–16.
6. Sarkar C, Sharma MC, Gaikwad S, Sharma C, Singh VP. Choroid plexus papilloma: a clinicopathological study of 23 cases. *Surg Neurol.* 1999;52(1):37–39.
7. Carter AB, Price Jr DL, Tucci KA, Lewis GK, Mewborne J, Singh HK. Choroid plexus carcinoma presenting as an intraparenchymal mass. *J Neurosurg.* 2001;95(6):1040–1044.
8. Kieserman S, Linstrom C, McCormick S, Petschenik AJ. Choroid plexus papilloma of the cerebellopontine angle. *Am J Otol.* 1996;17(1):119–122.
9. Sasani M, Solmaz B, Oktenoglu T, Ozer AF. An unusual location for a choroid plexus papilloma: the pineal region. *Childs Nerv Syst.* 2014;30(7):1307–1311.
10. Talacchi A, De Micheli E, Lombardo C, Turazzi S, Bricolo A. Choroid plexus papilloma of the cerebellopontine angle: a twelve patient series. *Surg Neurol.* 1999;51(6):621–629.
11. Wanibuchi M, Margraf RR, Fukushima T. Densely calcified atypical choroid plexus papilloma at the cerebellopontine angle in an adult. *J Neurol Surg Rep.* 2013;74(2):77–80.

12. Zhang TJ, Yue Q, Lui S, Wu QZ, Gong QY. MRI findings of choroid plexus tumors in the cerebellum. *Clin Imag*. 2011;35(1):64–67.
13. Wolff JE, Sajedi M, Brant R, Coppes MJ, Egeler RM. Choroid plexus tumours. *Br J Cancer*. 2002;87(10):1086–1091.
14. Coates TL, Hinshaw Jr DB, Peckman N, et al. Pediatric choroid plexus neoplasms: MR, CT, and pathologic correlation. *Radiology*. 1989;173(1):81–88.
15. Leblanc R, Bekhor S, Melanson D, Carpenter S. Diffuse craniospinal seeding from a benign fourth ventricle choroid plexus papilloma. Case report. *J Neurosurg*. 1998;88(4):757–760.
16. Sawaishi Y, Yano T, Yoshida Y, et al. Choroid plexus carcinoma presented with spinal dysfunction caused by a drop metastasis: a case report. *J Neurooncol*. 2003;63(1):75–79.
17. Yan C, Xu Y, Feng J, et al. Choroid plexus tumours: classification, MR imaging findings and pathological correlation. *J Med Imaging Radiat Oncol*. 2013;57(2):176–183.
18. Vandesteen L, Drier A, Galanaud D, et al. Imaging findings of intraventricular and ependymal lesions. *J Neuroradiol*. 2013;40(4):229–244.
19. Dubbioso R, Pappata S, Quarantelli M, et al. Atypical clinical and radiological presentation of cryptococcal choroid plexitis in an immunocompetent woman. *J Neurolo Sci*. 2013;334(1-2):180–182.
20. Falangola MF, Petito CK. Choroid plexus infection in cerebral toxoplasmosis in AIDS patients. *Neurology*. 1993;43(10):2035–2040.
21. Guermazi A, Miaux Y, Zagdanski AM, Laval-Jeantet M. Choroid plexitis caused by cytomegalovirus in a patient with AIDS. *Am J Neuroradiol*. 1996;17(7):1398–1399.
22. Hagiwara E, Nath J. Choroid plexitis in a case of systemic nocardiosis. *Emerg Radiol*. 2007;14(5):337–343.
23. Kovoor JM, Mahadevan A, Narayan JP, et al. Cryptococcal choroid plexitis as a mass lesion: MR imaging and histopathologic correlation. *Am J Neuroradiol*. 2002;23(2):273–276.
24. Krishnan S, Brown PD, Scheithauer BW, Ebersold MJ, Hammack JE, Buckner JC. Choroid plexus papillomas: a single institutional experience. *J Neurooncol*. 2004;68(1):49–55.
25. Kumari R, Raval M, Dhun A. Cryptococcal choroid plexitis: rare imaging findings of central nervous system cryptococcal infection in an immunocompetent individual. *Br J Radiol*. 2010;83(985):e14–e17.
26. Mongkolrattanothai K, Ramakrishnan S, Zagardo M, Gray B. Ventriculitis and choroid plexitis caused by multidrug-resistant Nocardia pseudobrasiliensis. *Pediatr Infect Dis*. 2008;27(7):666–668.
27. Gopal P, Parker JR, Debski R, Parker Jr JC. Choroid plexus carcinoma. *Arch Pathol Lab Med*. 2008;132(8):1350–1354.
28. McLendon RE, Rosenblum MK, Bigner DD, eds. *Russell and Rubinstein's Pathology of Tumors of the Nervous System*. London: CRC Press; 2006.
29. Chow E, Jenkins JJ, Burger PC, et al. Malignant evolution of choroid plexus papilloma. *Pediatr Neurosurg*. 1999;31(3):127–130.
30. Manjila S, Miller E, Awadallah A, Murakami S, Cohen ML, Cohen AR. Ossified choroid plexus papilloma of the fourth ventricle: elucidation of the mechanism of osteogenesis in benign brain tumors. *J Neurosurg*. 2013;12(1):13–20.

31. Centeno BA, Louis DN, Kupsky WJ, Preffer FI, Sobel RA. The AgNOR technique, PCNA immunohistochemistry, and DNA ploidy in the evaluation of choroid plexus biopsy specimens. *Am J Clin Pathol*. 1993;100(6): 690–696.

32. Anderson DR, Falcone S, Bruce JH, Mejidas AA, Post MJ. Radiologic-pathologic correlation. Congenital choroid plexus papillomas. *Am J Neuroradiol*. 1995;16(10):2072–2076.

33. Romano F, Bratta FG, Caruso G, et al. Prenatal diagnosis of choroid plexus papillomas of the lateral ventricle. A report of two cases. *Prenat Diagn*. 1996;16(6):567–571.

34. Safaee M, Oh MC, Bloch O, et al. Choroid plexus papillomas: advances in molecular biology and understanding of tumorigenesis. *Neuro Oncol*. 2013;15(3):255–267.

35. Safaee M, Oh MC, Sughrue ME, et al. The relative patient benefit of gross total resection in adult choroid plexus papillomas. *J Clin Neurosci*. 2013;20(6): 808–812.

36. Sun MZ, Ivan ME, Clark AJ, et al. Gross total resection improves overall survival in children with choroid plexus carcinoma. *J Neurooncol*. 2014;116(1): 179–185.

37. Chow E, Reardon DA, Shah AB, et al. Pediatric choroid plexus neoplasms. *Int J Radiat Oncol Biol Phys*. 1999;44(2):249–254.

38. Packer RJ, Perilongo G, Johnson D, et al. Choroid plexus carcinoma of childhood. *Cancer*. 1992;69(2):580–585.

39. Safaee M, Clark AJ, Bloch O, et al. Surgical outcomes in choroid plexus papillomas: an institutional experience. *J Neurooncol*. 2013;113(1):117–125.

40. McEvoy AW, Harding BN, Phipps KP, et al. Management of choroid plexus tumours in children: 20 years experience at a single neurosurgical centre. *Pediatr Neurosurg*. 2000;32(4):192–199.

41. Wrede B, Liu P, Ater J, Wolff JE. Second surgery and the prognosis of choroid plexus carcinoma – results of a meta-analysis of individual cases. *Anticancer Res*. 2005;25(6C):4429–4433.

42. Berger C, Thiesse P, Lellouch-Tubiana A, Kalifa C, Pierre-Kahn A, Bouffet E. Choroid plexus carcinomas in childhood: clinical features and prognostic factors. *Neurosurgery*. 1998;42(3):470–475.

43. Mazloom A, Wolff JE, Paulino AC. The impact of radiotherapy fields in the treatment of patients with choroid plexus carcinoma. *Int J Radiat Oncol Biol Phys*. 2010;78(1):79–84.

44. Allen J, Wisoff J, Helson L, Pearce J, Arenson E. Choroid plexus carcinoma – responses to chemotherapy alone in newly diagnosed young children. *J Neurooncol*. 1992;12(1):69–74.

45. Tabori U, Shlien A, Baskin B, et al. *TP53* alterations determine clinical subgroups and survival of patients with choroid plexus tumors. *J Clin Oncol*. 2010;28(12):1995–2001.

46. Wrede B, Liu P, Wolff JE. Chemotherapy improves the survival of patients with choroid plexus carcinoma: a meta-analysis of individual cases with choroid plexus tumors. *J Neurooncol*. 2007;85(3):345–351.

47. Berrak SG, Liu DD, Wrede B, Wolff JE. Which therapy works better in choroid plexus carcinomas? *J Neurooncol*. 2011;103(1):155–162.

48. Barman SL, Jean GW, Dinsfriend WM, Gerber DE. Choroid plexus papilloma – a case highlighting the challenges of extrapolating pediatric chemotherapy regimens to adult populations. *J Oncol Pharm Prac.* 2014.
49. Ogiwara H, Dipatri Jr AJ, Alden TD, Bowman RM, Tomita T. Choroid plexus tumors in pediatric patients. *Br J Neurosurg.* 2012;26(1):32–37.
50. Schreiber D, Bernstein K, Schneider J. Metastases of the central nervous system: a prospective study. 3rd Communication: metastases in the pituitary gland, pineal gland, and choroid plexus (author's transl). *Zentralbl Allg Pathol.* 1982;126(1–2):64–73.
51. Shapira Y, Hadelsberg UP, Kanner AA, Ram Z, Roth J. The ventricular system and choroid plexus as a primary site for renal cell carcinoma metastasis. *Acta Neurochir.* 2014;156(8):1469–1474.
52. Quinones-Hinojosa A, Chang EF, Khan SA, Lawton MT, McDermott MW. Renal cell carcinoma metastatic to the choroid mimicking intraventricular meningioma. *Can J Neurol Sci.* 2004;31(1):115–120.

Role of Blood–Brain Barrier, Choroid Plexus, and Cerebral Spinal Fluid in Extravasation and Colonization of Brain Metastases

Cecilia Choy*, Josh Neman[†]

*Division of Neurosurgery, City of Hope and Beckman Research Institute, Duarte, CA, USA; [†]Department of Neurosurgery, Keck School of Medicine, University of Southern California, Los Angeles, CA, USA

OUTLINE

The Choroid Plexus and Cerebrospinal Fluid. http://dx.doi.org/10.1016/B978-0-12-801740-1.00006-8

BRAIN METASTASES

Brain Metastases in the US Population

The most common cerebral tumors are metastases that form when cell from a primary site travels to the brain, forms brain metastases, and persists as a large problem in cancer – for patients' quality of life, physicians' clinical trials, and researchers' understanding of brain metastases itself. The 1-year survival rate in patients diagnosed with brain metastases remains less than 20%. A study published in 1985 reported a high frequency of brain metastases in comparison to primary brain tumors;[1] today, brain metastasis remains the most common tumor type in the brain. In the United States, about 170,000 cases of metastatic brain tumors are diagnosed yearly,[2] compared to about 23,000 cases annually of primary brain tumors.[3] The most frequent primary sites that metastasize to the brain are lung (40–50%), breast (15–25%), and melanoma (5–20%), while cancers such as sarcoma are unlikely have intracranial metastases.[4,5]

By the time brain metastasis is diagnosed, many patients present multiple tumor lesions. Although many symptoms displayed are contingent on the location of the metastasis, most can include progressive headaches, visual field disturbance, and seizures.[6] Responding to improved treatments for their primary tumors, patients are living longer, which increases their chances of developing and being diagnosed with brain metastases. The brain has also been suggested as a preferred site of cancer metastases for some primary tumor cells because of the "sanctuary" status of the brain, where tumor cells in the brain are shielded from treatments by the blood–brain barrier (BBB).[7] Currently in the clinic, contrast-enhanced magnetic resonance imaging (MRI) or computed tomography (CT) scan can be utilized to delineate brain lesions for brain metastases diagnosis. The contrast enhancement in these same imaging techniques can identify disruptions in the BBB, a large component of brain metastases to be discussed later in this chapter.

Much of brain metastases research has been developed quite recently, especially when compared to other areas of cancer research. The lack of research on brain metastases is due to improved therapies available for lung tumors, breast tumors, and melanomas,[8] in addition to a latent stage of cancer cell quiescence. Fewer than 0.02% of circulating tumor cells (CTCs) have the potential of successfully forming into metastases.[9] Yet, even with low metastatic potential of most cancer cells, it is suggested that some CTCs have the ability to establish metastasis quickly. *In vivo* data showed that a fibrosarcoma cell injected into an artery can establish itself in the cerebral space in 24 h, and in 5 days, can proliferate in peri-capillary spaces, which correlated with morphological changes in the endothelial cells at those time points.[10]

Fidler et al., through *in vivo* metastases experiments, observed that cancer cells from a certain metastasis showed increased ability to metastasize again in that organ.[11] It is still unclear why some cells have organotropism to the brain more than others. A cell best adapted to survive in a foreign microenvironment, for example, the brain, will be the cell that will flourish into metastasis according to Paget's "seed and soil" hypothesis.[12] Another hypothesis popular in the metastasis field was proposed by Ewing;[13] according to Ewing, tumor cells after dissemination circulate in the blood and metastasize to organs in proportion to the weight of and blood flow to the structures in the body.[4] Proteomic and genomic studies are starting to reveal commonalities between primary organ tumor samples from patients with brain metastases. It has been suggested through proteomic analysis that brain metastasis cells increased expression of enzymes involved in glucose oxidation, and their adaptation to produce minimal reactive oxygen species could promote resistance.[14] In brain metastases from breast cancers, cyclooxygenase 2 (COX2), an epidermal growth factor receptor ligand (HBEGF), and ST6GALNAC5 helped primary breast cancer cells migrate through the BBB to further enhance their metastatic capabilities.[15]

The Metastatic Cascade

The metastatic cascade explains the journey of a primary cancer cell that will spread to another tumor site. Primary cancer cells

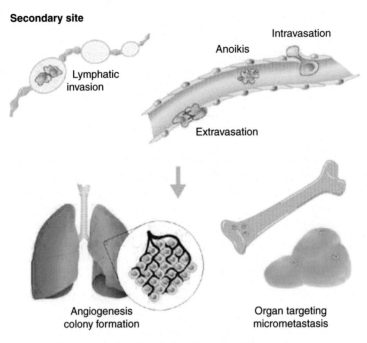

FIGURE 6.1   **The steps of a circulating tumor cell after arrest at a secondary tumor site during the end stages of metastasis.**[18]

can disseminate from the primary tumor, even at the start of tumor formation.[16] After a cell separates from the primary tumor and intravasates into the bloodstream, it circulates in the vasculature until it arrests at a secondary tumor site, extravasates in the organ parenchyma, and finally establishes micrometastasis before it becomes a full metastatic colony (Fig. 6.1).[17,18] For a cell to establish metastasis in any organ, the multiple steps of the cascade serve as barriers a cancer cell must overcome in order to metastasize. In the case of brain metastasis, a cell must overcome an additional barrier to invade the brain – the BBB.

There are two different forms of invasion – the "collective" form and the "amoeboid" form.[19] The collective form is found when cancer cells of a tumor "collectively" advance into the surrounding organ sites. The amoeboid form of invasion is frequently found in individual cancer cells that enter distant organs. The development of noninvasive *in vivo* techniques, the improvement of patient-derived xenograft models, and bioluminescence imaging have made it possible to image the spread of metastasis from primary site to metastasis organs.[20]

Extravasation

Extravasation in cancer cells is one of the final steps of the metastatic cascade, where the cancer cell traverses the endothelial cell layer after arresting in the vasculature in, for example, the brain, to enter the brain microenvironment. The site of metastasis and the origin of the primary tumor cells will both affect the rate of extravasation. For instance, small cell lung cancers frequently cross the BBB, an endothelial barrier, to establish metastasis in the brain. In one study, a cell took 48 h to extravasate into the brain, versus the 6 h to extravasate into the liver and 16 h to extravasate the lung.[21]

One theory of organotropism in the metastatic cascade is the mechanical entrapment theory, in which arrest at a distant organ is due to size limitations and dependent on the first capillary bed encountered by the disseminated tumor cell, but this theory was shown to be insufficient for tumor cell arrest.[22] In order to compensate for the large cell size, the cells can deform as when squeezed into narrow vessels, but a majority of cells cannot form successful colonies when investigated *in vivo*.[23] For this reason, cancer cells may be inhibited from proceeding in the capillaries. There is some controversy whether cancer cell extravasation is from mechanical entrapment theory or if it is due to the adhesion of endothelial cells of microvasculature, or both.[24] In order to establish its place in the brain microenvironment as a potential micrometastasis colony, a cancer cell may interact with endothelial cells through endothelial selectins, such as E-selectins and P-selectins, and glycoproteins followed by stable interactions with cell adhesion molecules (CAM), such as ICAM and VCAM.[24,25]

Extravasation also happens in normal physiology, demonstrated by the migration of leukocytes. Parallels have been determined between leukocyte extravasation and disseminated cancer cell extravasation utilizing *in vitro* cell culture studies in phase-contrast and time-lapse microscopy and transwell chambers.[26,27] Leukocytes establish extravasation through cytokine-activated transendothelial migration (TEM). Cytoskeletal changes aid in a leukocytes' extravasation process, including the steps of "docking" and "locking."[24] In leukocyte adhesion prior to complete extravasation, cytokines activate the endothelium, which triggers the expression of selectins. These selectins help dock the leukocyte, resulting in leukocyte rolling along the endothelial cells, mediated by a gradient of pro-inflammatory cytokines.[24]

However, leukocyte and tumor cell extravasation are not identical. Leukocytes are about 8 μm in diameter; cancer cells are much larger, up to twice as large as leukocytes and cannot pass through capillary vessels as easily as leukocytes.[24,28] Cancer cells need to somehow traverse the endothelial layer, which is much more difficult than for the smaller leukocytes. Disseminated tumor cells accomplish this by either utilizing growth factors like vascular endothelial growth factor (VEGF) and stromal cell-derived factor 1-alpha (SDF-1α), or by inducing apoptosis of endothelial cells to promote the retraction of the endothelial cell monolayer.[29,30] Furthermore, selectin-mediated rolling may or may not be required for tumor cell arrest, since leukocyte rolling is much more frequent than cancer cell rolling, although cancer cell rolling has been observed.[31] Cancer cells do express ligands for endothelial selectins, but unlike leukocytes, they are not used for the "docking" stage of extravasation.[32] After endothelial cell retraction, cancer cells adhere to the basement membrane to continue the metastatic cascade.

Colonization

The final step of the metastatic cascade is microcolonization followed by the formation of macrometastases, or metastasis colonies. Metastasis is described to be structured into two phases. First, there is tumor cell dissemination from the primary tumor site. Second, the cells adapt to a foreign microenvironment in order to successfully colonize into full metastasis.[19]

One uncertainty regarding metastases is the stage of tumor development at which the cancer cells are able to adapt to the foreign microenvironment. It is still not well understood whether these cancer cells have inherent adaptability because of the heterogenous nature of the cells in primary tumor formation or if the cancer cells develop the ability to adapt at the foreign tumor site as a result of selective pressure.[19] These cells can also manipulate the surrounding microenvironment for their metastatic advantage. Twenty-four hours after metastatic carcinoma cells were injected in the common carotid artery *in vivo*, there was vascularization as a result of co-option of blood vessels, and not as a result of angiogenesis, the formation of new blood vessels.[33] In establishing a metastatic colony, there is also support from the new foreign microenvironment for the disseminated cells, as they may receive help from surrounding stromal cells. More specifically, regarding the brain, astrocytes are the supportive cells around the tumor that may aid in the colonization of disseminated cells in the brain, among other forms of assistance to be discussed later in this chapter.[34,35]

Metastases to the Choroid Plexus

Although brain metastases is common, metastases to the choroid plexus is rare; but similar to other brain metastases, choroid plexus metastases is most commonly seen in adults. In some studies, they account for less than 1% of clinical brain metastases, compared to the higher levels of metastatic colonies found in intracranial, dural, and leptomeningeal regions discussed previously.[4,36] It is not established how tumors are spread to the choroid plexus, but it has been suggested that choroid plexus metastases develop when a disseminated tumor cell travels through the anterior/posterior choroidal arteries or when a disseminated tumor cell seeds in the cerebrospinal fluid (CSF).[36,37] In adults, choroid plexus metastases is most likely to be seen after primary colon, breast, bladder stomach, and thyroid cancer, while for the rare choroid plexus metastasis tumor in pediatric patients, neuroblastomas and retinoblastomas have been reported to be the source of the disseminated primary tumor cells.

BLOOD–BRAIN BARRIER VERSUS BLOOD–CSF BARRIER

The brain's fluid-based homeostasis is dependent on the complex functions of the BBB and the blood–CSF barrier (Fig. 6.2).[38,39] The CSF acts as a cushion from injury and a drain for any unwanted products from entering or lingering in the brain. In the early 1900s, Goldman showed that intravenously administered trypan blue dye could not reach the brain and spinal cord, but when the dye was injected into the CSF, the brain could be stained. From these results, he concluded that there was an impermeable barrier between the blood and the brain, but not between the brain and the CSF.[40]

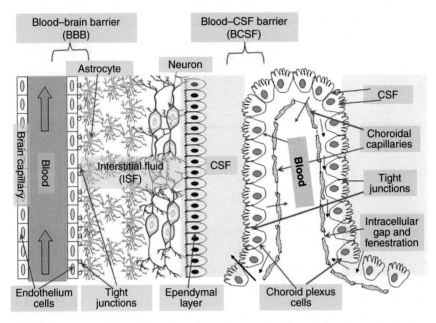

FIGURE 6.2   **The BBB and the blood–CSF barriers are the main barriers in the central nervous system.** These barriers help to protect the brain and maintain brain homeostasis.[39]

The Blood–Brain Barrier

The BBB helps maintain ionic and molecular homeostasis in the brain microenvironment through its main function of utilizing its capillary border to protect the brain from toxins, including pathogens and xenobiotics.[41] There are specific brain regions that lack a BBB, that is, the brain areas around the neurohypophysis, pineal glands, and areas of the brain regulating autonomic nervous system and endocrine glands; the BBB protects most of the brain by limiting entry into the brain.[42,43] The BBB controls the exchanges between the blood and the brain by utilizing paracellular, transcellular, enzymatic barriers, and efflux transporters.[5,42] First, they prevent passive diffusion of hydrophilic solutes. Substances trying to enter the brain, therefore, must move through the endothelial cell membranes. The ability to move through this membrane is largely determined by lipid solubility, although it is not the only determinant. Second, in addition to limiting passive diffusion, the BBB mediates the active transport of select nutrients into the brain. For example, amino acid transport is mediated through selective high affinity transport systems at the BBB.[44] Amino acid transport systems, such as System x^-, are sodium-independent transporters for L-glutamate and L-aspartate amino acids. System y^+ is a cationic amino acid transporter, and System L is a carrier of large neutral amino acids. Third, the BBB is involved in the efflux of hydrophobic molecules and drugs out of the brain back into the blood. This function can affect drug efficiency, which will be discussed later in this chapter. Last, the BBB regulates the TEM, or the extravasation, of circulating blood cells and pathogens.

To restrict the extravasation or intravasation of larger molecules, the BBB is largely made up of tight junctions between the capillary endothelial cells, which are only found in the brain. In addition to the tight junctions, there are adherens junctions and polarized transport systems. These transporters also aid in shuttling molecules, such as glucose. Molecules can also enter the brain by transient physical disruption of the BBB with the use of hyperosmotic agents, such as mannitol, which causes osmotic disruption of the BBB.[45] The tight junctions themselves must be regulated, which includes mediation through cyclic AMP (cAMP) and Rho GTPases and activation or degradation of occludins (regulatory proteins mediating paracellular permeability), and the zonula occludens proteins (cytoplasmic accessory proteins that help with organization of proteins at the plasma membrane).[42,43,46]

In addition to the endothelial cells that form tight junctions at the BBB, brain cells are involved in maintaining the structural integrity and function of the BBB. Astrocytic "end feet" are the terminal regions of astrocyte processes and they surround the capillary endothelial cells to help maintain the BBB. In addition to endothelial cells and astrocytes, there are pericytes at the basement membrane, potentially forming gap junctions, permitting ion exchange, and possibly regulating capillary diameter to control cerebral blood flow.[47] Microglia also come into contact with cerebral vessels, and although the exact function of their role is unclear, it has been suggested to have a protective function in ischemic brain injury response or potentiate damage in the BBB in brain inflammation.[48,49] Unfortunately, it is this same protective BBB that imposes limitations on brain uptake of cancer therapies.

The Blood–CSF Barrier

Another barrier of interest in the brain aside from the BBB is the blood–CSF barrier. The blood–CSF barrier is similar to the BBB in that it is a blood–brain interface, but the blood–CSF barrier is localized at the choroid plexus epithelium. The blood–CSF barrier controls the exchanges, as the name implies, between the blood and the CSF.[42] CSF is produced in and actively secreted by the choroid plexus, a vascular structure that exists in each of the ventricles of the brain. The capillaries of the choroid plexus are leaky, which allows macromolecules to pass through into the surrounding connective tissues.[50] Net transport of sodium and chloride across the epithelium will mediate the secretion of CSF.[51] Unlike the BBB, where endothelial cells form tight junctions, the blood–CSF barrier is made up of specialized epithelial cells. The choroid plexus' epithelial cells can also form adherence junctions, where adhesive contacts are formed from the interactions between cadherin and catenin proteins.[43]

The CSF system is a regulator of substance entry into the brain. The reasons for entry into the brain via blood–CSF barrier versus BBB are still not understood.[50] Like the BBB, the main function of the blood–CSF barrier is to inhibit free migration of molecules. The blood–CSF barrier is also involved in signal transduction and transportation regulation of hormones in the brain. However, there is no barrier to inhibit diffusion, even of larger molecules, from the CSF to the interstitial space of the nervous tissue.[52] It has been suggested that that the intracranial space and the vertebral canal should be perceived as the CSF space; this is considered separate from the brain and spinal cords' extracellular and intracellular space.[53]

Similarities and Differences Between BBB and Blood–CSF Barrier

Endothelial and epithelial tight junctions are made up of three major transmembrane protein families: (1) occludins, (2) claudins, and (3) junction-associated molecules (JAMs), in addition to cytoplasmic accessory proteins, which include the zonula occludens protein ZO-1.[42,43] Claudins are the major components of tight junctions. Matrix metalloproteinases (MMP), such as MMP-2 and MMP-9, were shown to mediate degradation of occludins and claudins to affect the integrity of the BBB.[54,55] Claudins on one cell bind homotypically to the claudins on the adjacent cell to form the seal of tight junction. The expression of claudin subtypes between endothelial tight junctions and epithelial tight junctions is similar but not identical. Epithelial cell tight junctions express claudin-1, claudin-2, and claudin-11, whereas endothelial cell tight junctions express claudin-1, claudin-3, claudin-5, and claudin-11.[42] Both types of tight junctions allow lipophilic substances to pass and block the passage of small hydrophilic substances and macromolecules from entering the brain and CSF without the assistance of specific carriers.

Tight junctions also have extra cytoplasmic accessory proteins, like cingulin and AF-6, two proteins that interact with the zonula occludens proteins; zonula occludens proteins are cytoplasmic accessory proteins that function as protein binding molecules to organize proteins at the plasma membrane.[56] The extracellular loops of occludins seal the junctions while their C-terminal region interacts with other proteins at the junction.[57] JAMs, involved in leukocyte extravasation, are localized at the tight junctions, and JAMs, such as JAM-C, are suggested to promote metastasis.[58]

Substance regulation, substance transport, and their movement from the blood into the CSF are similar to that of substances moving from the blood into the brain through the BBB – movement is accomplished through active transport systems or passive diffusion, but only for certain substances. The differences lie in the substances themselves being transported in the choroid plexus or the BBB. The BBB mediates much of the transport of oxygen, carbon dioxide, glucose, and amino acids.[51] The choroid plexus, on the other hand, is usually the entry site for compounds like calcium and some endocrine substances, such as insulin-like growth factor-II (IGF-2), and transport proteins like transthyretin.[50,51]

BBB Dysfunction in Cancer

To the advantage of disseminated cancer cells, the brain serves as a sanctuary site for the cancer cells can successfully extravasate into the brain, due to the inability of anticancer drugs to cross the BBB.[59] Although the brain is highly vascularized, the density of capillaries within the brain metastasis tumor is lower than the peritumoral region, the normal brain region surrounding the tumor.[60]

It is worth to note leptomeningeal metastasis, because it is one of the more popular brain metastasis sites at 8%. One way for tumors to spread to the leptomeninges is through a variety of routes, including through the choroid plexus.[61] Leptomeningeal metastasis develops in the microenvironment of the brain and the CSF; tumor cells increase their ability to metastasize by degrading the leptomeninges.[62]

Lessons Learned from Brain Injury

To date, more research has been dedicated to the effects of brain injury on the BBB and the response of the BBB to the brain injury than on the roles of the BBB in brain metastases. However, to the brain microenvironment, the stresses from brain injury and the stresses from a brain tumor are somewhat comparable. Therefore, we could apply the research done in brain injury to the studies done of the BBB and brain metastases.

Common consequences of traumatic brain injury include hematomas and axonal injury. Because the brain is made up of different structures with different properties, these different structures react differently in any one type of brain injury. And the structures react differently according to the brain injury itself, whether the injury results from a direct impact or acceleration–deceleration forces, such as the sudden movements from the forces of a car accident, when applied to the head. The impact and corresponding injury will disrupt the walls of the blood microvessels in the brain, activating a coagulation cascade and potentially increasing the permeability of the BBB. This changes the function of the BBB at the site, including the interaction with surrounding brain cells.

The BBB's normal function depends on the paracrine interactions between the brain endothelium and brain cells. The changes in BBB after brain injury are suggested to lessen neural tissue loss and also affect the efficacy of neuroprotective drugs.[63] After brain injury, alterations of some functional interactions between the glial cells and the endothelium, or the glial cells themselves, affect the functionality of the BBB. Astrocytes, the most numerous cell type in the brain, are a type of glial cell in the brain known to perform multiple supportive functions for normal brain homeostasis and contact the endothelium around the BBB's tight junction.[64] Microglia, another type of glial cell, suggested to be as the "macrophages" of the brain, are suggested to help normal BBB function through paracrine signaling, which has been described as the "gliovascular unit."[63,65] After brain injury, microglial cells monitor and maintain neuronal health, which includes a microglial cell's display of functional plasticity in morphology and receptor expression upon activation.

There are many factors that may contribute to BBB dysfunction after trauma, including transforming growth factor beta (TGF-β), which is stored primarily in astrocytes and microglia; TGF-β is up-regulated 6–12 h after injury.[66] Glutamate is released by astrocytes and is suggested to promote neuronal survival postinjury,[67] which contributes to astroglial swelling.[68] Another response to brain injury is reactive oxidative stress (ROS), which also affects the BBB function by increasing levels of hydroxyl radicals. This increase in radicals in turn may affect membrane lipids by resulting in an increase in the BBB permeability.[69] MMP, also a protein important in successful brain metastasis formation,[70] has a role in disrupting membrane integrity in the BBB because of its suggested degradation of the tight junction components.[71]

In addition to the response of the BBB, the blood–CSF barrier also plays a role in the regulation of the brain microenvironment after brain injury. After trauma to the brain, the blood–CSF barrier mediates leukocyte migration to the brain. The choroid plexus epithelia cells produce chemokines. These chemokines secreted by the choroid plexus, the vascularized tissue in which the blood–CSF barrier resides, is a requirement before leukocyte migration and may be a requirement for neutrophils and monocyte migration across the blood–CSF barrier.[72]

Blood–Brain Barrier and Brain Metastases

In vivo studies have been imperative in assessing BBB permeability and metastasis tumor size. Through real-time imaging, tumor cell migration and decrease of the BBB integrity have been observed.[73] The observations between tumor cells and BBB integrity have been at times and to date, somewhat irreconcilable. For example, there are tumor studies suggesting that increased VEGF results in leaky and permeable blood vessels, indicating BBB permeabilization.[74] But in other *in vivo* studies, there was no permeability found in the brain with diffuse tumor cells that did not form a large colony.[75]

Because of the conflicting results on the maintenance of the BBB integrity, whether maintaining the integrity of BBB in brain metastases is factor dependent, stage dependent, or cell dependent (to name a couple possibilities), the result of extravasation on BBB integrity is still unclear. But it is clear that shortly after extravasation, tumor cells maintain close contact with the endothelium in order to exploit protective functions of the BBB on the tumor cells; astrocytes, whose end feet interact with the endothelium of the BBB, have a protective effect on the tumor cells. Astrocytes can provide protection from chemotherapy by inducing upregulation of survival genes, such as *TWIST1*, in tumor cells; the infiltration of the tumor by reactive astrocytes had an additive protective function by providing a chemoprotective effect when altering the intracellular microenvironment, such as sequestering calcium from tumor cell cytoplasm to evade tumor cell apoptosis.[76,77] Although astrocytes are initially involved in the maintenance of normal brain microenvironment homeostasis, reactive astrocytes later during brain injury can protect the injured neurons from apoptosis.[78,79] In metastasis formation, these astrocytes can actually protect the brain metastases from cytotoxicity from chemotherapeutic agents.[80] Another reason for the prolonged contact between tumor and endothelium may be the support of the vascular basement membrane or vascular co-option through cancer-cell-released serpins protecting L1 cell adhesion molecule (L1CAM)-mediated vascular co-option for brain metastasis cell growth.[81,82]

BBB in Extravasation and Brain Colonization

The BBB is typically a large hurdle for cell trafficking; however, some metastatic cells are suggested to prefer metastasizing to the brain than any other organs.[11] There are multiple hypotheses regarding how a tumor cell disrupts the BBB to gain entry into the brain and complete the steps of the metastatic cascade; many of them focus on the permeability of BBB. These hypotheses include angiogenesis; by increasing levels of angiogenic growth factors, there is an increase in the disruption and permeability of the BBB function.[42]

The BBB has been suggested in some instances, to have a metastasis-friendly role, because as a part of the brain microenvironment, it can provide an environment that can increase the chances of a metastatic cell to successfully establish a metastatic colony. The cerebral endothelial cells could help the cells in transmigration (similar to leukocyte TEM) and help provide a friendly microenvironment for the metastatic cells.[5] TEM is similar to extravasation in the suggested steps of rolling, adhesion, and transmigration, but there are physiological, molecular, and mechanical differences between the immune cells that go through TEM and the metatastatic cells that extravasate.[5] Morphologically, tumor cells that extravasate out of the brain vasculature are elongated at arrest but become round to stretch the vessel walls for transmigration, at which time the cells narrow at the vascular wall.[83] The endothelial cells of the BBB retract, allowing tumor cells to enter the brain.[21] Leukocytes and tumor cells can go through TEM through two ways – transcellular migration and paracellular migration. The transcellular pathway, involves travel through the endothelial cell.[5] The other migration pathway, the paracellular method of migration, involves travel through interendothelial junctions. If a tumor cell is to transverse the BBB through paracellular transmigration, the cell requires disruption of the endothelial tight junctions and increased permeability.

SUCCESSFUL DRUG DELIVERY THROUGH THE BBB

A diagnosis of brain metastases results in a large impact on a patient and their caregivers' quality of life. Current initial treatment for patients diagnosed with brain metastases depends on several factors. These factors include the characteristics of the brain lesions themselves, such as number, size, and location. There are also patient-based characteristics, such as symptoms, the status of the metastases, and availability and preferences of treatment on behalf of the patients regarding treatment options.[84]

For symptomatic lesions, the patients typically undergo surgery and/or radiation. Treatment options can include surgical resection, whole brain radiation therapy (WBRT), radiosurgery, or a combination of these treatments. In order to minimize morbidity and mortality and to improve a patient's quality of life, it is necessary to either improve existing therapies or develop new therapies. Many have suggested improving combination strategies of existing therapies, including radiotherapy to the surgical bed and stereotactic radiosurgery even in addition to WBRT; stereotactic radiosurgery alone is worrisome in that its effectiveness is limited by tumor cavity size, which could advocate further usage of the whole brain radiation therapy.[85] Another suggestion is to further advance salvage treatment, also known as rescue treatment, where additional treatment is administered if a tumor does not respond to standard therapies. This includes advances in WBRT or stereotactic radiosurgery provided as salvage therapy and understanding the effects of neurocognitive function and tumor recurrence after salvage therapy.

There are many challenges in the treatment of brain metastases, including traversal of the BBB and unknown characteristics of the tumor itself. The therapeutic challenge in brain metastases treatment regarding the heterogeneity of the tumor makes it difficult to treat the whole brain metastasis lesion, even if the therapy can cross the BBB. One way to target metastasis tumor cells is to target the circulating tumor cells (CTCs) before they arrest at the brain; this requires understanding the gene or protein expression of circulating tumor cells that are more likely to metastasize to the brain. EpCAM$^-$ CTCs from primary breast tumor selective for brain metastases from patients had a consistent protein signature, including Her2$^+$, EGFR$^+$, HPSE$^+$ (heparanase), and Notch1$^+$, suggesting these markers could be utilized to predict cancer metastasis to the brain.[86] The heterogeneity in this signature was discovered at the gene level. And as mentioned previously, other studies utilizing gene-expression profiling of different patient samples with brain infiltrating cancer cells identified COX-2, HBEGF (EGFR ligand heparin-binding EGF-like growth factor), and ST6GalNAc5 as mediators for cancer cells penetrating the BBB to further establish metastasis in the brain microenvironment.[15] Signatures such as these can help identify and target circulating cells that would metastasize to the brain before they arrive at the brain; once they enter the brain microenvironment, some argue that it may be more difficult to target the tumor cells.

Common Problems Traversing the BBB

One of the main problems with treating brain metastases is development of targeted drugs that cross the BBB and remain in the brain. Hurdles regarding BBB traversal, efflux pumps, BBB integrity, and imaging continue to plague the development of better treatments. There are certain properties in drug development to ensure or at least increase potentials of safe passage into the brain. Without drug traversal of the BBB, the tumor remains in the "sanctuary" of the brain. Because of the difficulty in getting drugs into the brain, there is a need to transiently disrupt the BBB function and the physical barrier itself in order enhance the delivery of antitumor drugs, viral vectors, nanoparticles, or any other treatment options.[87,88]

A second difficulty in treating brain metastases is that drugs that cross the BBB must conform certain characteristics to traverse the BBB, whether it is traversal through passive diffusion or active transport. These characteristics are similar to those of treatments that must enter the CSF and extracellular space. These include limitations of molecular size, lipophilicity, and active transport by transporter proteins such as organic anion transporters (Oats).[53] Plasma protein binding is also a limiting factor because wherever there is an intact barrier, the unbound plasma fraction can cross the barrier since binding proteins pass the barrier to some degree.[89]

For molecules that are too large or too lipophilic to traverse the endothelial cell layer naturally, efflux transporters can help transport molecules across the BBB. One of these transporters is the ATP-binding cassette (ABC) family that transport the lipophilic molecules through a concentration gradient by ATP hydrolysis. One problem with these transporters is that transporters such as P-glycoprotein (P-gp) and other multidrug resistance proteins (MDR) efflux drugs are initially meant to target the brain but can later be pumped back out into the bloodstream.[90,91] Crizotinib, a substrate for the ATP-binding cassette drug efflux pump family member ABCG2 and P-gp, is an anaplastic lymphoma kinase tyrosine kinase inhibitor.[80,92] Crizotinib in combination with Elacridar, an inhibitor of the efflux pumps, increased the concentration of drug in the brain in an *in vivo* model. Not only is it imperative to create drugs that can get into the brain microenvironment, there is also a necessity in blocking the transporter activity of drugs out of the brain to improve treatment efficiency.

A third problem to treat brain metastases is the uncertainty of the levels of BBB damage at the site of brain metastases.

Observations of gadolinium-enhanced MRI scans have suggested that the very BBB integrity that limits drug uptake into the brain is incompetent in most brain metastases.[80] There are reports that in a murine model of breast cancer brain metastasis, intravascular tumor expansion was the cause of BBB disruption.[93] But there is also ongoing debate that after the tumor cells damage the endothelial vessel wall, there is a possibility that the BBB repairs the damage after tumor cell extravasation; in addition, there is also debate that the BBB actually maintains its integrity during metastases.[5,94,95]

A fourth barrier in the development of improved therapies presents after the therapy is developed and is ready for use as treatment. There are still hurdles regarding imaging to diagnose brain metastases and predict drug permeability. In terms of diagnosis, earlier a metastatic lesion is detected, more promising are the results of treatment. MRI is typically utilized to diagnose brain metastasis. However, because of the difficulty in identifying and distinguishing a solitary metastasis versus a primary tumor, such as a high grade glioma, the method of detection also needs improvement. Recent advances in MRI imaging have helped differentiate between metastatic and primary tumors.[80] In order to improve detection of metastatic brain lesions, *in vivo* models are employed. A method using tumor necrosis factor (TNF) and to a lesser extent, lymphotoxin, has been developed to selectively permeabilize BBB at metastasis sites from breast cancer, which allowed smaller tumor detection and the increase of trastuzumab (Herceptin) delivery to breast cancer brain metastasis lesions.[96] Magnetic resonance spectroscopy (MRS) is an improvement over the usual MRI technique because it differentiates parts of the brain based on the levels of specific metabolites. For example, cancers have increased levels of choline and decreased levels of neuronal biomarker N-acetylaspartate (NAA).[97] NAA can be found in primary and metastatic tumors themselves, but where NAA levels differ between the two tumor types is in the peritumoral regions, the normal area of brain adjacent to the tumor. The ratio of choline to NAA in the peritumoral region of a glioma is higher than normal brain, while in metastases, the peritumoral regions have about the same choline and NAA levels as the normal brain. Relative cerebral blood volume (rCBV) in a high grade glioma is also higher in the peritumoral region than the normal brain, while the rCBV levels in brain metastases' peritumoral brain's levels are comparable to normal brain.

Requirements for a Successful BBB-Traversing Drug

There is no current standard cytotoxic chemotherapy regimen for patients with brain metastases; in fact, sometimes they are treated with chemotherapies typically used for extracranial tumors.[80] Yet, in order to improve the existing or to develop new treatments, it is not only necessary to understand the tumor itself, but also the surrounding brain microenvironment, with special emphasis on the BBB. The new treatments must be able to cross the BBB to target the micrometastases lesions.

One solution to target brain metastases is to target the cells that have a signature of brain metastases before the cell arrests at the brain site and traverses the BBB into the brain. Molecular characterization will improve tumor targeting. It was shown that molecular drivers may lead to increased risk of cancer, such as Her2[+] tumors.[98] They are determined through the protein-expression profile studies, the signature Her2[+], EGFR[+], HPSE[+] (heparanase), and Notch1[+] cells may be a targetable signatures before CTCs successfully extravasate into the brain.[86]

Another targetable area in successfully treating brain metastases is to target the interactions between tumor cell and endothelial cells immediately before and after extravasation. As previously mentioned, there is a prolonged interaction between the tumor cell and the BBB after the tumor cell extravasates, indicating that the endothelial cells could initially provide an advantage for survival for these tumor cells in the new microenvironment. One way to target these interactions between tumor cells and the BBB is to target and inhibit the signaling pathways that promote extravasation out of the blood vessel and/or increase the tumor cell survival in the brain.

A third option in improving targeted brain metastasis treatment would be to increase receptor-mediated transport of drugs into the brain. As is with receptors that mediate transcytosis of their ligand through the BBB, drugs could exploit the same characteristic to transport drugs into the brain microenvironment. Currently, clinical trials in treatments utilizing the influx of drug transporters are rare. But exploiting drug transporters for therapy is not impossible; clinical trials include inhibiting the receptor for advanced glycation end products (RAGE) and platelet/endothelial cell adhesion molecule-1 (PECAM-1), which is able to mediate peptide influx into the brain.[40]

Another targetable aspect of brain metastases is to target the mechanisms that brain metastasis cell can manipulate in order to fully establish a metastatic colony. These mechanisms include signaling utilized for migration or ones utilized for proliferation once in the brain microenvironment. One mechanism that cells utilize is the Rho/Rac signaling pathway for an "amoeboid"-type migration, much like the migration type of leukocytes. Rho-promoted signaling or elongated cell motility through Rac-dependent signaling increases actomyosin contractile forces to remodel the cytoskeleton into a "rounded bleb" to aid in cell migration.[99] Other targets are proteins involved in Src signaling whose activation typically results in disruption of the BBB. Src signaling is known to alter the interendothelial junctions through the protein interactions between the zona occludin proteins and the Src-family tyrosine kinase inhibitor, PP2. This interaction results in increased tyrosine phosphorylation in occludins, which disrupts BBB tight junction integrity.[100]

For cells that have successfully entered the brain microenvironment, therapies can be developed to target mechanisms that a metastasis cell uses to establish macrometastasis. In order to survive in the brain and form a metastatic colony, tumor cells must proliferate despite any defense mechanisms that may be provided by the brain against the tumor cells. One pathway, that is, involved in the survival and proliferation of tumor cells is the phosphoinositide-3-kinase (PI3K) pathway. The activation of downstream pathway proteins regulates survival and proliferation of brain metastases from breast cancers and melanoma.[101,102] The inhibition of these pathways could lead to tumor cell apoptosis, or at the very least, stall or inhibit tumor growth.

Because there are many hurdles a brain metastasis must overcome, there are, in turn, many targets of brain metastases that can be developed for treatment (Fig. 6.3).[103] The increased popularity in brain metastases research is fairly recent. Through *in vitro* and *in vivo* studies that further reveal the tendency for certain primary cancer cells to travel to the brain, in addition to identifying mechanisms brain metastasis cells utilize to establish lesions and the provisions from the brain microenvironment that aid in establishing metastases, better treatments can be developed.

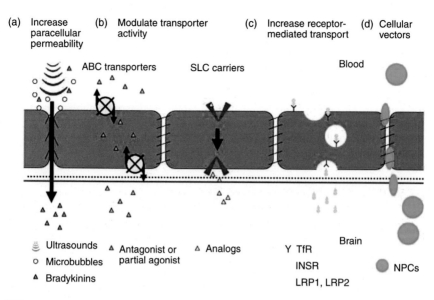

FIGURE 6.3 **There are many opportunities in drug development to circumvent the problems presented by the BBB in drug delivery to the brain.**[103]

References

1. Walker AE, Robins M, Weinfeld FD. Epidemiology of brain tumors: the national survey of intracranial neoplasms. *Neurology*. 1985;35(2):219–226.
2. Platta CS, Khuntia D, Mehta MP, Suh JH. Current treatment strategies for brain metastasis and complications from therapeutic techniques: a review of current literature. *Am J Clin Oncol*. 2010;33(4):398–407.
3. American Cancer Society. *Cancer Facts and Figures*; 2014.
4. Posner JB, Chernik NL. Intracranial metastases from systemic cancer. *Adv Neurol*. 1978;19:579–592.
5. Wilhelm I, Molnar J, Fazakas C, Hasko J, Krizbai IA. Role of the blood–brain barrier in the formation of brain metastases. *Int J Mol Sci*. 2013;14(1):1383–1411.
6. Lassman AB, DeAngelis LM. Brain metastases. *Neurol Clin*. 2003;21(1):1–23:vii.
7. Palmieri D, Chambers AF, Felding-Habermann B, Huang S, Steeg PS. The biology of metastasis to a sanctuary site. *Clin Cancer Res*. 2007;13(6):1656–1662.
8. Chiang AC, Massague J. Molecular basis of metastasis. *N Engl J Med*. 2008;359(26):2814–2823.
9. Luzzi KJ, MacDonald IC, Schmidt EE, et al. Multistep nature of metastatic inefficiency: dormancy of solitary cells after successful extravasation and limited survival of early micrometastases. *Am J Pathol*. 1998;153(3):865–873.
10. Ballinger Jr WE, Schimpff RD. An experimental model for cerebral metastasis: preliminary light and ultrastructural studies. *J Neuropathol Exp Neurol*. 1979;38(1):19–34.

11. Fidler IJ. Selection of successive tumour lines for metastasis. *Nat New Biol.* 1973;242(118):148–149.
12. Paget S. The distribution of secondary growths in cancer of the breast. 1889. *Cancer Metastasis Rev.* 1989;8(2):98–101.
13. *Neoplastic Diseases: A Treatise on Tumors* [computer program]. Philadelphia: W.B. Saunders; 1928.
14. Chen EI, Hewel J, Krueger JS, et al. Adaptation of energy metabolism in breast cancer brain metastases. *Cancer Res.* 2007;67(4):1472–1486.
15. Bos PD, Zhang XH, Nadal C, et al. Genes that mediate breast cancer metastasis to the brain. *Nature.* 2009;459(7249):1005–1009.
16. Eyles J, Puaux A-L, Wang X, et al. Tumor cells disseminate early, but immunosurveillance limits metastatic outgrowth, in a mouse model of melanoma. *J Clin Invest.* 2010;120(6):2030–2039.
17. Talmadge JE, Fidler IJ. AACR centennial series: the biology of cancer metastasis: historical perspective. *Cancer Res.* 2010;70(14):5649–5669.
18. Tapon N, Ziebold U. Invasion and metastasis: stem cells, screens and survival. *EMBO Rep.* 2008;9(11):1078–1083.
19. Hanahan D, Weinberg RA. Hallmarks of cancer: the next generation. *Cell.* 2011;144(5):646–674.
20. Lu X, Kang Y. Organotropism of breast cancer metastasis. *J Mammary Gland Biol Neoplasia.* 2007;12(2–3):153–162.
21. Paku S, Dome B, Toth R, Timar J. Organ-specificity of the extravasation process: an ultrastructural study. *Clin Exp Metastasis.* 2000;18(6):481–492.
22. Glinskii OV, Huxley VH, Glinsky GV, Pienta KJ, Raz A, Glinsky VV. Mechanical entrapment is insufficient and intercellular adhesion is essential for metastatic cell arrest in distant organs. *Neoplasia.* 2005;7(5):522–527.
23. Barbera-Guillem E, Smith I, Weiss L. Cancer-cell traffic in the liver. I. Growth kinetics of cancer cells after portal-vein delivery. *Int J Cancer.* 1992;52(6):974–977.
24. Miles FL, Pruitt FL, van Golen KL, Cooper CR. Stepping out of the flow: capillary extravasation in cancer metastasis. *Clin Exp Metastasis.* 2008;25(4):305–324.
25. Steinbach F, Tanabe K, Alexander J, et al. The influence of cytokines on the adhesion of renal cancer cells to endothelium. *J Urol.* 1996;155(2):743–748.
26. Kramer RH, Nicolson GL. Interactions of tumor cells with vascular endothelial cell monolayers: a model for metastatic invasion. *Proc Nat Acad Sci USA.* 1979;76(11):5704–5708.
27. Li YH, Zhu C. A modified Boyden chamber assay for tumor cell transendothelial migration *in vitro. Clin Exp Metastasis.* 1999;17(5):423–429.
28. Chambers AF, MacDonald IC, Schmidt EE, Morris VL, Groom AC. Clinical targets for anti-metastasis therapy. *Adv Cancer Res.* 2000;79:91–121.
29. Kebers F, Lewalle JM, Desreux J, et al. Induction of endothelial cell apoptosis by solid tumor cells. *Exp Cell Res.* 1998;240(2):197–205.
30. Fischer S, Clauss M, Wiesnet M, Renz D, Schaper W, Karliczek GF. Hypoxia induces permeability in brain microvessel endothelial cells via VEGF and NO. *Am J Physiol.* 1999;276(4):C812–820.
31. Giavazzi R, Foppolo M, Dossi R, Remuzzi A. Rolling and adhesion of human tumor cells on vascular endothelium under physiological flow conditions. *J Clin Invest.* 1993;92(6):3038–3044.

32. Satoh M, Numahata K, Kawamura S, Saito S, Orikasa S. Lack of selectin-dependent adhesion in prostate cancer cells expressing sialyl Le(x). *Int J Urol.* 1998;5(1):86–91.

33. Holash J, Maisonpierre PC, Compton D, et al. Vessel cooption, regression, and growth in tumors mediated by angiopoietins and VEGF. *Science.* 1999;284(5422):1994–1998.

34. Wang L, Cossette SM, Rarick KR, et al. Astrocytes directly influence tumor cell invasion and metastasis *in vivo*. *PloS One.* 2013;8(12).

35. Xing F, Kobayashi A, Okuda H, et al. Reactive astrocytes promote the metastatic growth of breast cancer stem-like cells by activating Notch signalling in brain. *EMBO Mol Med.* 2013;5(3):384–396.

36. Qasho R, Tommaso V, Rocchi G, Simi U, Delfini R. Choroid plexus metastasis from carcinoma of the bladder: case report and review of the literature. *J Neurooncol.* 1999;45(3):237–240.

37. Kohno M, Matsutani M, Sasaki T, Takakura K. Solitary metastasis to the choroid plexus of the lateral ventricle. Report of three cases and a review of the literature. *J Neurooncol.* 1996;27(1):47–52.

38. Segal MB. The choroid plexuses and the barriers between the blood and the cerebrospinal fluid. *Cell Mol Neurobiol.* 2000;20(2):183–196.

39. Bhaskar S, Tian F, Stoeger T, et al. Multifunctional Nanocarriers for diagnostics, drug delivery and targeted treatment across blood-brain barrier: perspectives on tracking and neuroimaging. *Part Fibre Toxicol.* 2010;7:3.

40. Zlokovic BV. The blood–brain barrier in health and chronic neurodegenerative disorders. *Neuron.* 2008;57(2):178–201.

41. Purves DAG, Fitzpatrick D, et al. *Neuroscience.* 2nd ed. Sunderland, MA: Sinauer Associates; 2001.

42. Weiss N, Miller F, Cazaubon S, Couraud PO. The blood–brain barrier in brain homeostasis and neurological diseases. *Biochim Biophys Acta.* 2009;1788(4): 842–857.

43. Ballabh P, Braun A, Nedergaard M. The blood–brain barrier: an overview: structure, regulation, and clinical implications. *Neurobiol Dis.* 2004;16(1):1–13.

44. Smith QR. Transport of glutamate and other amino acids at the blood–brain barrier. *J Nutr.* 2000;130(4S suppl):1016–1022.

45. Joshi S, Ergin A, Wang M, et al. Inconsistent blood brain barrier disruption by intraarterial mannitol in rabbits: implications for chemotherapy. *J Neurooncol.* 2011;104(1):11–19.

46. Hirase T, Staddon JM, Saitou M, et al. Occludin as a possible determinant of tight junction permeability in endothelial cells. *J Cell Sci.* 1997;110(14):1603–1613:1997.

47. Cuevas P, Gutierrez-Diaz JA, Reimers D, Dujovny M, Diaz FG, Ausman JI. Pericyte endothelial gap junctions in human cerebral capillaries. *Anat Embryol.* 1984;170(2):155–159.

48. Denes A, Vidyasagar R, Feng J, et al. Proliferating resident microglia after focal cerebral ischaemia in mice. *J Cereb Blood Flow Metab.* 2007;27(12): 1941–1953.

49. Nishioku T, Dohgu S, Takata F, et al. Detachment of brain pericytes from the basal lamina is involved in disruption of the blood–brain barrier caused by lipopolysaccharide-induced sepsis in mice. *Cell Mol Neurobiol.* 2009;29(3):309–316.

50. Nilsson C, Lindvall-Axelsson M, Owman C. Neuroendocrine regulatory mechanisms in the choroid plexus–cerebrospinal fluid system. *Brain Res Brain Res Rev.* 1992;17(2):109–138.
51. Laterra J, Keep R, Betz L, Goldstein G. *Basic Neurochemistry: Molecular, Cellular, and Medical Aspects.* 6th ed. Philadelphia: Lippincott-Raven; 1999.
52. Reese TS, Feder N, Brightman MW. Electron microscopic study of the blood–brain and blood–cerebrospinal fluid barriers with microperoxidase. *J Neuropathol Exp Neurol.* 1971;30(1):137–138.
53. Nau R, Sorgel F, Eiffert H. Penetration of drugs through the blood–cerebrospinal fluid/blood–brain barrier for treatment of central nervous system infections. *Clin Microbiol Rev.* 2010;23(4):858–883.
54. Chiu P-S, Lai S-C. Matrix metalloproteinase-9 leads to claudin-5 degradation via the NF-κB pathway in BALB/c mice with eosinophilic meningoencephalitis caused by *Angiostrongylus cantonensis. PloS One.* 2013;8(3):e53370.
55. Liu J, Jin X, Liu KJ, Liu W. Matrix metalloproteinase-2-mediated occludin degradation and caveolin-1-mediated claudin-5 redistribution contribute to blood-brain barrier damage in early ischemic stroke stage. *J Neurosci.* 2012;32(9):3044–3057.
56. Hawkins BT, Davis TP. The blood–brain barrier/neurovascular unit in health and disease. *Pharmacol Rev.* 2005;57(2):173–185.
57. Wong V, Gumbiner BM. A synthetic peptide corresponding to the extracellular domain of occludin perturbs the tight junction permeability barrier. *J Cell Biol.* 1997;136(2):399–409.
58. Langer HF, Orlova VV, Xie C, et al. A novel function of junctional adhesion molecule-C in mediating melanoma cell metastasis. *Cancer Res.* 2011;71(12):4096–4105.
59. Steeg PS. Tumor metastasis: mechanistic insights and clinical challenges. *Nat Med.* 2006;12(8):895–904.
60. Leenders W, Kusters B, Pikkemaat J, et al. Vascular endothelial growth factor-A determines detectability of experimental melanoma brain metastasis in GD-DTPA-enhanced MRI. *Int J Cancer.* 2003;105(4):437–443.
61. Steeg PS, Camphausen KA, Smith QR. Brain metastases as preventive and therapeutic targets. *Nat Rev Cancer.* 2011;11(5):352–363.
62. Pedersen PH, Rucklidge GJ, Mork SJ, et al. Leptomeningeal tissue: a barrier against brain tumor cell invasion. *J Natl Cancer Inst.* 1994;86(21):1593–1599.
63. Chodobski A, Zink BJ, Szmydynger-Chodobska J. Blood-brain barrier pathophysiology in traumatic brain injury. *Translat Stroke Res.* 2011;2(4):492–516.
64. Chen Y, Swanson RA. Astrocytes and brain injury. *J Cereb Blood Flow Metab.* 2003;23(2):137–149.
65. Streit WJ. Microglial response to brain injury: a brief synopsis. *Toxicol Pathol.* 2000;28(1):28–30.
66. Cook JL, Marcheselli V, Alam J, Deininger PL, Bazan NG. Temporal changes in gene expression following cryogenic rat brain injury. *Brain Res Mol Brain Res.* 1998;55(1):9–19.
67. Ikonomidou C, Turski L. Why did NMDA receptor antagonists fail clinical trials for stroke and traumatic brain injury? *Lancet Neurol.* 2002;1(6):383–386.
68. Maxwell WL, Bullock R, Landholt H, Fujisawa H. Massive astrocytic swelling in response to extracellular glutamate – a possible mechanism for post-traumatic brain swelling? *Acta Neurochir Suppl Wien.* 1994;60:465–467.

69. Mertsch K, Blasig I, Grune T. 4-Hydroxynonenal impairs the permeability of an *in vitro* rat blood–brain barrier. *Neurosci Lett*. 2001;314(3):135–138.

70. Zohrabian VM, Nandu H, Gulati N, et al. Gene expression profiling of metastatic brain cancer. *Oncol Rep*. 2007;18(2):321–328.

71. Yang Y, Estrada EY, Thompson JF, Liu W, Rosenberg GA. Matrix metalloproteinase-mediated disruption of tight junction proteins in cerebral vessels is reversed by synthetic matrix metalloproteinase inhibitor in focal ischemia in rat. *J Cereb Blood Flow Metab*. 2007;27(4):697–709.

72. Szmydynger-Chodobska J, Strazielle N, Zink BJ, Ghersi-Egea JF, Chodobski A. The role of the choroid plexus in neutrophil invasion after traumatic brain injury. *J Cereb Blood Flow Metab*. 2009;29(9):1503–1516.

73. Fazakas C, Wilhelm I, Nagyőszi P, et al. Transmigration of melanoma cells through the blood–brain barrier: role of endothelial tight junctions and melanoma-released serine proteases. *PloS One*. 2011;6(6):e20758.

74. Lee TH, Avraham HK, Jiang S, Avraham S. Vascular endothelial growth factor modulates the transendothelial migration of MDA-MB-231 breast cancer cells through regulation of brain microvascular endothelial cell permeability. *J Biol Chem*. 2003;278(7):5277–5284.

75. Zhang RD, Price JE, Fujimaki T, Bucana CD, Fidler IJ. Differential permeability of the blood-brain barrier in experimental brain metastases produced by human neoplasms implanted into nude mice. *Am J Pathol*. 1992;141(5):1115–1124.

76. Kim SJ, Kim JS, Park ES, et al. Astrocytes upregulate survival genes in tumor cells and induce protection from chemotherapy. *Neoplasia*. 2011;13(3):286–298.

77. Lin Q, Balasubramanian K, Fan D, et al. Reactive astrocytes protect melanoma cells from chemotherapy by sequestering intracellular calcium through gap junction communication channels. *Neoplasia*. 2010;12(9):748–754.

78. Fidler IJ, Balasubramanian K, Lin Q, Kim SW, Kim SJ. The brain microenvironment and cancer metastasis. *Mol Cells*. 2010;30(2):93–98.

79. Mahesh VB, Dhandapani KM, Brann DW. Role of astrocytes in reproduction and neuroprotection. *Mol Cell Endocrinol*. 2006;246(1–2):1–9.

80. Owonikoko TK, Arbiser J, Zelnak A, et al. Current approaches to the treatment of metastatic brain tumours. *Nat Rev Clin Oncol*. 2014;11(4):203–222.

81. Carbonell WS, Ansorge O, Sibson N, Muschel R. The vascular basement membrane as "soil" in brain metastasis. *PloS One*. 2009;4(6):e5857.

82. Valiente M, Obenauf AnnaC, Jin X, et al. Serpins promote cancer cell survival and vascular co-option in brain metastasis. *Cell*. 2014;156(5):1002–1016.

83. Kienast Y, von Baumgarten L, Fuhrmann M, et al. Real-time imaging reveals the single steps of brain metastasis formation. *Nat Med*. 2010;16(1):116–122.

84. Lin NU. Breast cancer brain metastases: new directions in systemic therapy. *Ecancermedicalscience*. 2013;7:307.

85. Goetz P, Ebinu JO, Roberge D, Zadeh G. Current standards in the management of cerebral metastases. *Int J Surg Oncol*. 2012;:493426:2012.

86. Zhang L, Ridgway LD, Wetzel MD, et al. The identification and characterization of breast cancer CTCs competent for brain metastasis. *Sci Transl Med*. 2013;5(180):180ra148.

87. Kroll RA, Neuwelt EA. Outwitting the blood–brain barrier for therapeutic purposes: osmotic opening and other means. *Neurosurgery*. 1998;42(5):1083–1099.

88. Muldoon LL, Nilaver G, Kroll RA, et al. Comparison of intracerebral inoculation and osmotic blood–brain barrier disruption for delivery of adenovirus, herpesvirus, and iron oxide particles to normal rat brain. *Am J Pathol*. 1995;147(6):1840–1851.

89. Norrby SR. Role of cephalosporins in the treatment of bacterial meningitis in adults. Overview with special emphasis on ceftazidime. *Am J Med*. 1985;79(2A):56–61.

90. Decleves X, Amiel A, Delattre JY, Scherrmann JM. Role of ABC transporters in the chemoresistance of human gliomas. *Curr Cancer Drug Targets*. 2006;6(5):433–445.

91. Schinkel AH, Mayer U, Wagenaar E, et al. Normal viability and altered pharmacokinetics in mice lacking mdr1-type (drug-transporting) P-glycoproteins. *Proc Natl Acad Sci USA*. 1997;94(8):4028–4033.

92. Costa DB, Kobayashi S, Pandya SS, et al. CSF concentration of the anaplastic lymphoma kinase inhibitor crizotinib. *J Clin Oncol*. 2011;29(15):443–445.

93. Lu W, Bucana CD, Schroit AJ. Pathogenesis and vascular integrity of breast cancer brain metastasis. *Int J Cancer*. 2007;120(5):1023–1026.

94. Fazakas C, Wilhelm I, Nagyoszi P, et al. Transmigration of melanoma cells through the blood-brain barrier: role of endothelial tight junctions and melanoma-released serine proteases. *PloS One*. 2011;6(6):e20758.

95. Lorger M, Felding-Habermann B. Capturing changes in the brain microenvironment during initial steps of breast cancer brain metastasis. *Am J Pathol*. 2010;176(6):2958–2971.

96. Connell JJ, Chatain G, Cornelissen B, et al. Selective permeabilization of the blood–brain barrier at sites of metastasis. *J Natl Cancer Inst*. 2013;105(21): 1634–1643.

97. Law M, Cha S, Knopp EA, Johnson G, Arnett J, Litt AW. High-grade gliomas and solitary metastases: differentiation by using perfusion and proton spectroscopic MR imaging. *Radiology*. 2002;222(3):715–721.

98. Kallioniemi OP, Holli K, Visakorpi T, Koivula T, Helin HH, Isola JJ. Association of c-erbB-2 protein over-expression with high rate of cell proliferation, increased risk of visceral metastasis and poor long-term survival in breast cancer. *Int J Cancer*. 1991;49(5):650–655.

99. Sahai E, Marshall CJ. Differing modes of tumour cell invasion have distinct requirements for Rho/ROCK signalling and extracellular proteolysis. *Nat Cell Biol*. 2003;5(8):711–719.

100. Takenaga Y, Takagi N, Murotomi K, Tanonaka K, Takeo S. Inhibition of Src activity decreases tyrosine phosphorylation of occludin in brain capillaries and attenuates increase in permeability of the blood-brain barrier after transient focal cerebral ischemia. *J Cereb Blood Flow Metab*. 2009;29(6):1099–1108.

101. Nanni P, Nicoletti G, Palladini A, et al. Multiorgan metastasis of human HER-2+ breast cancer in Rag2−/−;Il2rg−/− mice and treatment with PI3K inhibitor. *PloS One*. 2012;7(6).

102. Davies MA, Stemke-Hale K, Lin E, et al. Integrated molecular and clinical analysis of AKT activation in metastatic melanoma. *Clin Cancer Res*. 2009;15(24):7538–7546.

103. Weiss N, Miller F, Cazaubon S, Couraud P-O. The blood–brain barrier in brain homeostasis and neurological diseases. *Biochim Biophys Acta*. 2009;1788(4):842–857.

The Role of the Choroid Plexus in the Pathogenesis of Multiple Sclerosis

Vahan Martirosian, Alex Julian, Josh Neman

Department of Neurological Surgery, Keck School of Medicine,
University of Southern California, Los Angeles, CA, USA

OUTLINE

The Choroid Plexus and Cerebrospinal Fluid. http://dx.doi.org/10.1016/B978-0-12-801740-1.00007-X

THE NEURAL IMMUNE SYSTEM COMPARED WITH THE PERIPHERAL IMMUNE SYSTEM

The anatomical divisions that compose the central nervous system (CNS) require a stable and nurturing environment, devoid of the oscillations and fluctuations experienced by peripheral organs. The reason for this is because of the vulnerability of the sensitive nervous tissue and its importance to critical functions of livelihood. Perturbations in this parenchyma could cause dire effects, leading to abnormal physical function. One method in protecting this integrity in a healthy individual is by separating the innate immune system of the CNS from that of the peripheral organs. This function is mediated through the blood–brain barrier (BBB), which stonewalls any immune cell or molecules, from entering the neural parenchyma. Although this may seem counter-intuitive as pathogens always seem to appear everywhere, regardless of any bodily defense mechanisms, the CNS maintains this in order to protect itself from neuroinflammation during onset of illness. This symptom is caused by the peripheral immune system, which is severely threatening if not more dangerous than the disease itself. This notion is supported by studies that have shown that inflammation of neural tissue causes impairment in brain function.[1-3]

In order to handle pathogenic invasion, the brain's innate immune cells, microglia and astrocytes, play an important role in quickly recognizing subtle changes in the microenvironment and responding quickly by recruiting CNS-specific macrophages and T-cells for pathogen clearance (Fig. 7.1). However, the response in the periphery of the body differs from that of the brain's immune reaction. In the periphery, dendritic cells internalize antigens from various infectious particles, process them internally, and present them to naïve T-cells through the MHC class II receptor. This initiates the maturation and subsequent differentiation of the naïve T-cells into T_h1, T_h2, or T_h17 T-cells, which each secrete cytokines and chemokines that proceed to stimulate repair mechanisms and recruit more immune cells. In the brain, the presence of infectious particles induces activation of glial cells, especially astrocytes and microglia. These in turn secrete chemokines and cytokines, which recruit T-cells into the neural parenchyma. Presentation of the antigen causes the differentiation into the same subtypes as the peripheral immune system.[4] However, it is important to note that the

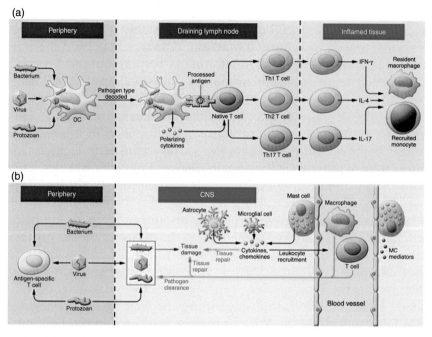

FIGURE 7.1   **Innate immunity in the periphery and CNS.** (a) In the face of a peripheral infection, innate immune cells prime and instruct T lymphocytes. Tissue DCs internalize microbial protein antigens, process them into peptides, and display them on their surfaces with MHC class II molecules; migrate to draining lymph nodes; and present antigens to naive CD4+ T cells. DCs direct the quality of the subsequent inflammatory response by decoding distinctive pathogen-associated signals and transmitting this information to T cells in the form of regulatory cytokines such as IL-12 (for Th1), IL-4 (Th2), or IL-6/TGF-β (Th17). In addition, lymph node environmental cues can provide information about the site of infection (gut, skin, or other). Armed with this information, effector T cells migrate to infected tissues. Upon reactivation, Th1, Th2, and Th17 cells express phenotype-defining cytokines that act on resident and recruited innate cells, which operate collectively with factors such as complement to clear the infection. (b) Resident microglia and astrocytes exert multiple functions in the CNS, including protective and restorative responses to CNS infection or injury. Cytokines and chemokines expressed by resident CNS cells also promote the recruitment of circulating lymphocytes and myeloid cells from the periphery to assist in pathogen clearance. Innate responses in the CNS cannot directly initiate adaptive immunity. Innate CNS reactions also occur during neuroinflammatory disorders and utilize many of the same components as do host defense responses. *Reprinted with permission from.[4]*

differentiated cells that were recruited to the brain most likely had a stable ratio between T_h1 and T_h2, as an unstable ratio of these cells is shown to lead to abnormal function.[5] The reason behind this balance is that T_h2 reduces the inflammatory response of the T_h1 cells, and further recruits noninflammatory macrophages, such as healing M2 monocyte-derived macrophages, to the site of injury, thus limiting the derogatory effects of inflammation in the brain.[6,7] This phenomenon occurs in the presence of dysfunction or disease that results in the accumulation of active T-cells in the neural sphere. Interestingly, while this is the case in diseased patients, studies have shown that in a healthy brain, T-cells are completely absent.[8] This may be a result of T-cells being removed from the environment, as it has been shown that when a small amount of antigens are present, naïve T-cells undergo "deletion" in the periphery.[9] This process may also occur in the brain when naïve T-cells randomly infiltrate for immune surveillance. The low concentration of antigens would then render the T-cell useless, causing them to be removed or "deleted" from the brain, thus explaining the lack of T-cell presence in a healthy individual. We can further postulate that when antigens are present in high amounts, neural T-cells become activated and mature leukocytes from the periphery are recruited. Thus, the brain protects itself from the immune system by causing removal of T-cells from its parenchyma when they are not needed. The understanding of these fundamental concepts are integral to the remaining sections in this chapter as we further discuss the specific mechanisms related to the regulation of the immune system in the presence of disease.

THE IMPORTANCE OF THE PRESENCE OF IMMUNE CELLS IN THE NEURAL PARENCHYMA

Evolution has ensured an atmosphere absent of immunity in the brain in order to reduce the chance of inflammation. Thus, the brain has become known as one of the few immune-privileged sites in the body. However, studies in the past few decades have begun to demonstrate that our previous perception of the presence of immune cells in the brain being detrimental may not be entirely true. Research has shown that immunity plays a vital role in maintenance of the neuronal architecture and regulation of learning, memory, and behavior.[8] Without this system present in the neural parenchyma, brain plasticity and function deteriorated, evidenced by studies that examined that immune-deficient mice were impaired in hippocampal function, which interrogates spatial memory.[10] More specifically speaking, CD4$^+$ T-cells are directly involved in the preservation of the delicate environment of the brain. They were shown to encode specific brain self-antigens, deemed autoimmunity, which has been shown to inhibit neuroinflammation. The intricate balance between the activity of the innate CNS and adaptive immune systems plays an important role in reducing neuroinflammation and boosting autoimmunity, as shown in studies where injection of CNS antigens caused protection of the neural parenchyma in various neurodegenerative diseases.[11] Importantly, direct T-cell interactions were not detectable in a normal brain, thus leading to the belief that the support provided by these circulating immune cells was either indirect or provided extraneous of the neural environment.[12,13] Therefore, we can conclude that the presence of the immune system in the brain may not only be for defense mechanisms, but also to secrete nurturing factors for the growing and developing brain.

THE CHOROID PLEXUS AS A POSSIBLE HOT SPOT FOR IMMUNE CELL INFILTRATION

Current studies have shed light on possible routes of entry for these circulating immune cells, indicating the choroid plexus (CP) as a hot spot for infiltration. Previously, it was envisioned that the main route of neural access for the peripheral immune system was through the BBB. To fully grasp how these mechanisms are mediated by the CP, it is important to understand its location and anatomy. The choroid plexus is a structure located in the ventricles of the brain that functions in producing cerebrospinal fluid. The CP is vascularized, as it consists of epithelial cells that surround the blood vessels.[14] The structure of the CP resembles that of the BBB; however, the CP incorporates a fenestrated endothelium that has a high permeability to the blood circulation incased in a tight junction epithelium. The CP is centrally located in the neural parenchyma, situated in the ventricles of the brain.

More focus is being placed on other functions of the CP, besides its role in CSF production. These include its mediation in cell trafficking as well as its role in the preservation of the neural parenchyma. The infiltration of recruited healing M2 macrophages, as mentioned earlier, was shown to be recruited to the site of neural injury without the breakdown of the BBB.[6] Further studies showed that the cells enter through the blood–CSF barrier, which provided evidence and shed light on the role of the CP in the infiltration of these cells.[6,15] In a normal brain, most of the T-cells were found to be in the CP, the cerebrospinal fluid (CSF), and the meningeal membranes covering the brain.[16] CD4[+] T cells have been shown to either enter the CP through the fenestrated epithelium of the CP or maintain themselves in the CP (Fig. 7.2). The presence of T-cells in the CP was shown to regulate immune cell trafficking across the blood–CSF border through the production of interferon-γ (IFN-γ), which subsequently upregulates adhesion molecules such as ICAM-1 and VCAM-1. This mechanism holds true during physiological conditions as well as CNS damaged conditions.[6,15] Studies show that mice devoid of IFN-γ had reduced numbers of immune cells in their brains.[17] Thus, these observations allow us to conclude that IFN-γ and the CP are used for the trafficking of immune cells during normal surveillance of the neural parenchyma. Moreover, this also explains the presence of the immune cells in the CSF as well as the meningeal membranes in the absence of disease.

FIGURE 7.2   **Innate Immunity in the Periphery and CNS.** (a) In the face of a peripheral infection, innate immune cells prime and instruct T lymphocytes. Tissue DCs internalize microbial protein antigens, process them into peptides, and display them on their surfaces with MHC class II molecules; migrate to draining lymph nodes; and present antigens to naive CD4[+] T cells. DCs direct the quality of the subsequent inflammatory response by decoding distinctive pathogen-associated signals and transmitting this information to T cells in the form of regulatory cytokines such as IL-12 (for Th1), IL-4 (Th2), or IL-6/TGF-β (Th17). In addition, lymph node environmental cues can provide information about the site of infection (gut, skin, or other). Armed with this information, effector T cells migrate to infected tissues. Upon reactivation, Th1, Th2, and Th17 cells express phenotype-defining cytokines that act on resident and recruited innate cells, which operate collectively with factors such as complement to clear the infection. (b) Resident microglia and astrocytes exert multiple functions in the CNS, including protective and restorative responses to CNS infection or injury. Cytokines and chemokines expressed by resident CNS cells also promote the recruitment of circulating lymphocytes and myeloid cells from the periphery to assist in pathogen clearance. Innate responses in the CNS cannot directly initiate adaptive immunity. Innate CNS reactions also occur during neuroinflammatory disorders and utilize many of the same components as do host defense responses. *Reprinted with permission from.[8]*

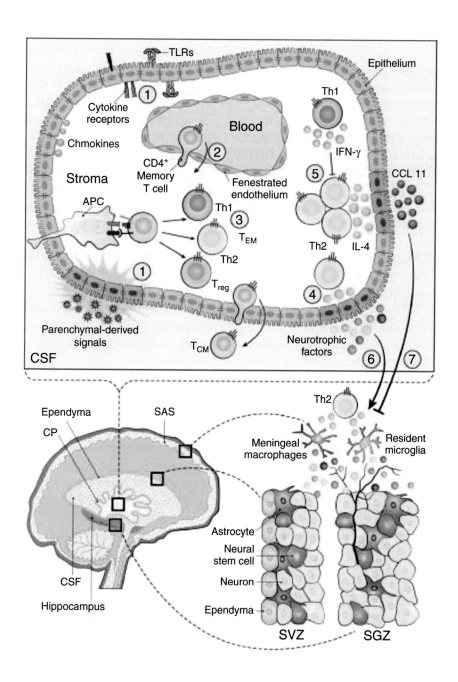

Furthermore, the anatomical location of the CP provides further evidence that it may be involved or at least harbor the cells that aid in maintaining brain plasticity and neurogenesis. The ependymal cell layer, which hosts neural stem cells, and subventricular and subgranual zones (SVZ, SGZ, respectively), which mediates neurogenesis, are all closely associated with the CP. The CP also secretes growth factors and other peptides into the CSF all of which benefit the neural parenchyma.[18] For example, factors such as brain-derived neurotrophic factor (BDNF) and insulin-like growth factor (IGF-1) are released by the CP under physiological and diseased conditions.[19] Again, the upregulation of these neurotrophic factors was shown to be associated with the presence of immune cells harbored in the CP. In fact, it was shown that T_h1 and T_h2 cells secrete the factors that aid in the preservation of the neural environment. Interestingly, although these cells maintain themselves in the CP, they remain in an inactive state until antigen presenting cells, which ingest antigens from the CSF, reactivate them and induce productions of cytokines, chemokines, and neurotrophic factors (Fig. 7.2). The fact that these cells remain in a dormant state may be due to the brain's inherent ability to control the autoimmune ability of these cells and thus protect itself from autoimmune disorders. Studies have shown that if certain cytokines, like CCL11, begin to inhibit adult neurogenesis and secretion of neurotrophic factors by T_h2 cells, these shifts the immune cells into an inflammatory state, which may provide clues on the propagation of inflammation in the brain in autoimmune disorders.[8]

In light of its anatomy, the CP receives signals and is affected by both sides of its domain, the brain and the rest of the body. Specifically, the CP contains receptors, such as Toll-like receptors, which identify distress signals from the brain and subsequently cause a proinflammatory response by the CP epithelium (Fig. 7.2). In conjunction with this, Shechter et al. has demonstrated that the CP and its specific receptors still responds to CNS injuries even though it is not in the proximal area. They showed that an acute injury to the CNS spinal cord cause the CP to recruit and allow transport of anti-inflammatory macrophages, which are necessary for repair. The signal cascade began from the site of injury, which released cytokines into the CSF, which was then recognized by the CP and caused infiltration of immune cells into the CSF and to the injury site. Likewise, the CP can also respond to immune signals in the blood circulation. Some dangerous changes, such as ion imbalance or chronic inflammation that occur in the peripheral organs, may eventually negatively affect the brain.[20,21] Thus, the CP protects the neural microenvironment by changing its transcriptome to maintain a balanced atmosphere. Therefore, we can clearly assume the role the CP has in mediating and controlling the immune response in the CNS.

In summary, the CNS immune response is regulated, especially by the CP. Most of this is done through an autoimmune mechanism that recruits specific noninflammatory macrophages to the injury site in the CNS for reparation. However, what if these pathways were disrupted by age-related deficiencies in the CP epithelial layer or by neurodegenerative disorders like multiple sclerosis (MS)? How would these issues change the CP and what would be the result of this? Could simple disruptions in the production of autoantigens by neural cells exploit the CNS-specific antigens of the innate T-cells? Moving forward, we will provide answers for these questions and focus specifically on (1) how MS changes the neuronal architecture and the CP and (2) possible treatments mediated by immunotherapy.

BACKGROUND ON MULTIPLE SCLEROSIS

MS is a disease that causes demyelination of nerve cells in the spinal cord and brain. The symptoms include motor dysfunction, but may also consist of mental, emotional, and psychiatric problems.[22,23] MS has been postulated to be caused by genetics, environmental factors, or by infectious particles, such as microbes or viruses.[22,24] However, although these factors are important in the subsequent pathogenesis of MS, they may serve more to explain how the inflammatory response is initiated rather than explain why infiltrating cells attack specific targets in the CNS. They also do not explain how T-cells, which normally have no direct contact with the neuronal architecture of the brain, as mentioned previously, suddenly gain the ability to attack and destroy the brain. Thus, as we will explain later, there may be other occurrences that lead these T-cells to become cytotoxic and therefore initiate MS pathology.

While it has long been known that MS is an autoimmune disease that affects the white matter of the brain,[22] recent studies have discovered that demyelination of gray matter, specifically the deep gray matter (DGM), also occurs extensively in this disease.[25] This is extremely important as these lesions are specifically seen in progressive MS and may play an important role in the physical and cognitive deficits seen in patients. Moreover, researchers have recently determined that in most MS cases, the caudate nucleus and the thalamus, both DGM structures, are severely affected. In some of the analyzed tissue sections, the caudate nucleus was found to be almost completely demyelinated. This is extremely important as the caudate nucleus is largely responsible for voluntary movement, which is severely affected in MS. Furthermore, it was also observed that most of the lesions were along the ventricular surface. Interestingly, these lesions are seen in even the earliest stages of the development of the disease. The study was also able to discern a correlation between lesions in the DGM and cortex, while such a correlation was not found between the white matter tracts and the gray matter. It was also observed that DGM lesions were usually a mix of white and gray matter lesions, but in some specific cases, the lesions were only restricted to the GM, while the WM was spared. Last, astrocytes were only observed in WM matter lesions, while they were rarely seen in the GM.[25]

Taking these data as a whole, one can notice the two separate methods in which lesions are caused in MS. The first one encapsulates the role of CSF, and inherently the CP, in the onset of lesions in multiple sclerosis, especially in the early stages of the disease. The second pathway incorporates the classical paradigm of BBB-mediated entry and initialization of pathogenesis. In the next section, we will address each of these pathways specifically.

THE RELATIONSHIP BETWEEN THE CHOROID PLEXUS AND MULTIPLE SCLEROSIS PATHOGENESIS

Before the establishment of the role of the CP in MS, most studies focused on how MS initiated through the BBB. Although this still serves as a possible method of action in the pathogenesis of MS, it seems more plausible to hypothesize that the CP is more responsible for the immune infiltration seen in MS. As mentioned previously, the structure of the CP differs from that of the BBB. Due to the fenestrated epithelium, this inherently causes more infiltration from the blood vessel into the CP and thus makes the blood–CSF barrier leakier than the BBB. This easier method of infiltration should mean that most of the immune cells infiltrate through the CP and thus aggregate in the CSF in the lateral ventricles, the anatomical location of the CP. This was also seen in a previous study by,[26] in which inflammation of the CP occurred before inflammation in the brain, providing more evidence that infiltration through the CP occurs earlier than infiltration through the BBB. Furthermore, accumulation of lymphocytes in this region may serve to explain why DGM lesions were found to be in the earlier stages of MS. Because of the proximity of the DGM to the ventricles and the now lymphocyte-infested CSF, it serves as a prime target of destruction for the cytotoxic function of the T-cells. Moreover, this early inflammation can also be attributed to the immediate immune response to antigens in the CSF, thus indicating the more efficient transfer of cytokines to the peripheral blood and subsequent recruitment of lymphocytes into the neural parenchyma.[27]

On the other hand, lesion formation away from the ventricles and CSF most likely occurs through a mechanism involving the BBB. Because the BBB is not as leaky as the blood–CSF barrier, the infiltration of the leukocytes and subsequent initiation of lesions occurs at a later time point from that of the lesions formed by CP-infiltrated immune cells. Furthermore, the difference in initiating factors of the lesions may be the reason as to why no correlation between gray and white matter lesions was seen. White matter, which is more central in the brain, may be exposed to lymphocytes and macrophages through a compromised BBB. This break in the impenetrable lining of the BBB is caused by the expression of a variety of cell surface adhesion markers, which initiate the migration of lymphocytes across the BBB. Therefore, it may be that infiltration of lymphocytes through the CP induces GM/DGM lesions, while lymphocytes from the BBB cause the WM lesions. This also provides reasoning as to why glial cells like astrocytes were mostly observed in the WM. Astrocytes mediate the entry of the lymphocytes through cytokines and other factors that influence the BBB to become permeable. Hence, the astrocytes would be most likely to be found in WM lesions, where they would facilitate the migration of lymphocytes into the neural parenchyma.

PHYSICAL CHANGES THAT OCCUR IN THE CHOROID PLEXUS AS A RESULT OF IMMUNE CELL INFILTRATION

Moving forward, we will focus on the first pathway of immune infiltration (through the CP), and highlight changes in the molecular make-up of the CP to further examine its role in the onset of

MS. In one study,[28] the analysis of several MS brains indicated high quantities of HLA-DR immunostained cells around the blood vessels inside the CP. HLA-DR, which is part of the human leukocyte antigen subclass, is an MHC class II cell surface receptor that is ciphered by the human leukocyte antigen complex located on chromosome 6. The DR represents the two subunits in the heterodimer that forms the functional receptor. Interestingly, a genetic defect in this complex on chromosome 6 represents one of the possible causes of multiple sclerosis, as mentioned earlier. Furthermore, HLA-DR staining was observed on the CP epithelial cells as well as on the epiplexus cells, which are CP-associated cells on the CSF side of the membrane. In contrast, normal brain tissue did not express HLA-DR positive cells in the epiplexus cells, and was only seen in extremely rare situations in the CP epithelial cells. It should be noted that because this staining was absent in healthy tissue does not necessarily mean that there was an absence of T-cells in the CP. The absence of these cells may indicate that they are simply not active, as studies have shown the presence of naïve T-cells in the stroma of the CP.[28]

Second, high levels of CD68 staining in the CP stroma as well as in between the CP epithelial cells were observed. CD68 is a glycoprotein that is expressed by immune cells like lymphocytes and macrophages. CD68 was rarely stained in the CP in control brains. Further double-staining with CD68 and HLA-DR was carried out to confirm that the stained cells expressed both markers. The combination of the previous two points allows us to postulate that in an MS brain, there are differentiated lymphocytes actively seeking antigens. Third, T-cells were also identified in the MS brain, using CD3 staining. Following suit with the rest of the markers, T-cells were again absent in normal brains. Finally, VCAM-1 staining was carried out, indicating that the CP vessels were stained but the CP epithelium was not. VCAM-1 is a cell adhesion marker, which mediates the adhesion of immune cells to the endothelium. Again, VCAM-1 was not observed in healthy brains. Once again, this assertion agrees with a previous statement made where differentiation of naïve T-cells into their Th1 subtype, and subsequent release of IFN-γ causes upregulation of VCAM-1. Moreover, this only occurs in diseased states and under normal CNS surveillance. However, it should be important to consider that T-cells, even in normal conditions, would probably only infiltrate the brain if specific pathogenic antigens were provided to the innate naïve T-cells (Fig. 7.3).

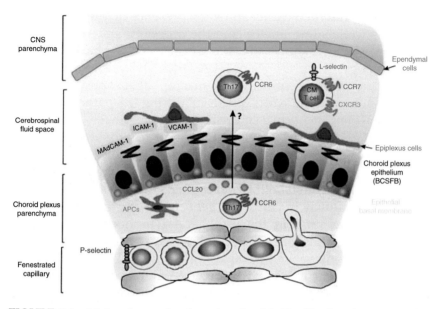

FIGURE 7.3 Molecular mechanisms involved in T cell migration across the epithelial blood–cerebrospinal fluid barrier (BCSFB). The choroid plexus might be a preferential T-cell entry site into the CNS during immunosurveillance, that is, in the absence of neuroinflammation. Circulating T cells extravasate in a P-select independent manner across fenestrated capillaries to reach the choroid plexus parenchyma, which is outside the CNS. To reach the CSF-filled ventricles, T cells need to breach the BCSFB established by choroid plexus epithelial cells. The paracellular pathway is sealed by unique tight junctions between the choroid plexus epithelial cells. CCR6+ T helper (Th)17 cells may use chemokine CC ligand (CCL)20 expressed by choroid plexus epithelium as a guidance cue to migrate across the BCSFB into the CSF filled ventricular space. The high number of central memory T (TCM) cells found in the CSF of humans suggests that this T cell subset preferentially crosses the BCSFB. The molecular mechanisms used by TCM cells to migrate across the BCSFB are unknown. Functional expression of intercellular adhesion molecule (ICAM)-1, vascular cell adhesion molecule (VCAM)-1, and under inflammatory conditions, mucosal addressin cell adhesion molecule (MAdCAM)-1 is restricted to the apical surface of choroid plexus epithelial cells, and is thus not available for the basolateral to apical migration of immune cells across the BCSFB. The choroid plexus is in constant movement and these adhesion molecules might instead allow T cells to crawl along the surface of the choroid plexus epithelium or alternatively might mediate the adhesion of antigen-presenting cells, so called epiplexus cells, ensuring their strategic localization behind the BCSFB to present CNS antigens to the T cells during immunosurveillance. *Reprinted with permission from.*[29]

The data presented earlier show the presence of immune cells in the CP as well as immune cells in the process of entering the CSF during MS, which shows that the CP in fact does have a role in MS pathology. The presence of HLA-DR/CD68 positive cells may indicate the presence of activated macrophages, and thus shows the presence of immune cells in the CP during MS pathology while they are absent in normal brain tissue. Furthermore, the intense staining of HLA-DR in the CP stroma indicates that the CP is undergoing activation, which causes the epithelial cells, and thus the associated epiplexus cells, which are known as phagocytotic cells, to begin ingesting metabolites and proteins that are in the CSF. This may be the pathway in which antigens in the CSF are detected and eventually presented to the peripheral lymphocytes, indicating that these cells may have possible antigen presenting cell (APC) characteristics. Moreover, the staining of macrophages in between epithelial cells may represent the process of infiltration by these immune cells from the blood to the CSF. The presence of VCAM-1 and activated T-cells, and the inherent absence of these cells in normal brains, further supplements the hypothesis of CP-mediated infiltration of immune cells in MS. Studies of the CSF composition have shown increased numbers of activated T-cells, memory B-cells, and antibody secreting cells. Antagonizing the function of VCAM-1 reduces the number of these cells, thus supporting the role of this molecule in immune cell infiltration.[28]

We previously introduced one mechanism in which recruitment of proinflammatory macrophages can occur through APCs traveling from the CSF to the peripheral lymph system and presenting antigens to the immune cells.[30] However, other theories exist that explain other possible mechanisms in which activation of these macrophages and subsequent MS pathogenesis can occur. Another theory states that naïve pioneer T-cells, which serve the purpose of random monitoring of the neural environment, enter the CSF and travel to the subarachnoid space where APCs activate them.[31] This differs from the previous theory as the other study hypothesized that the APCs initially bind to the antigens then travel to the peripheral blood where they activate the immune cells.

While these two mechanisms signify the transport of antigens from the CSF to the periphery, and subsequent recruitment of immune cells and onset of MS, others hypothesize that chronic inflammation in the periphery may result in changes to the choroid plexus and thus serve as the initial step in pathogenesis. Studies have shown that chronic inflammation as opposed to acute inflammation is more likely to result in a CNS disease.[21] The pathway of pathogenesis is regulated through sustained gene expression changes in the choroid plexus. In one particular study, lipopolysaccharide (LPS), known to cause strong immune responses in animals, was injected into mice once every 2 weeks, for 3 months to emulate a sustained inflammation state. The mice were sacrificed and analyzed for gene expression 3 and 15 days after the last injection of LPS. Interestingly, upon analysis of the choroid plexus, the 3 days postinjection animals had 536 probes with altered expression, while only 46 probes were found to be altered in animals sacrificed 3 days after a single dose of LPS. This clearly shows the large-scale change caused by chronic inflammation in the CP. Moreover, chronic treatment with LPS induced sustained transcription of genes that were already found in mice injected with a single dose. Thus, chronic treatment causes a sustained if not permanent change in the gene expression of the CP. Furthermore, 15 days after a single injection of LPS, animals returned to basal levels of gene expression while animals given a chronic dose, still had altered levels of gene expression after 15 days, albeit them being lower than the animals sacrificed 3 days after the last injection. This shows that a chronic inflammatory response causes sustained changes in the gene expression of the CP and inhibits the return to basal levels of gene expression. Importantly, of the genes that remained altered 15 days postinjection, many were related to leukocyte migration and the complement cascade signaling pathways. The altered genes mediate the production of cytokines and chemokines responsible for trafficking of immune cells from the blood into tissues, which cause the recruitment of monocytes, T-cells, and dendritic cells. Moreover, these genes are also involved in the production of cell adhesion molecules such as ICAM-1 and MadCAM-1. Interestingly, the production of these cytokines and cell adhesion molecules did not show to lead to increased numbers of immune cells in the CSF.[21]

When considering both hypotheses of immune cells migration into the neural parenchyma, it would be logical to consider that the entire process may be a combination of the two. Chronic inflammatory stimulation induces gene expression changes in the CP, causing chemokine production and subsequent immune cell recruitment. Production of cell adhesion molecules then allows the docking of the immune cells. Epiplexus cells, acting as APCs, ingest the antigens from the CSF, and travel into the CP and initiate the infiltration of the bound immune cells. This pathway, along with the idea of the high permeability of the CP fenestrated epithelium, may be a possible mechanism in the recruitment of immune cells into the brain. Propagation of these inflammatory pathways, as mentioned previously, could be initiated by the presence of viruses or other infectious particles. The mechanism behind these initiations lies in the ability of these molecules to secrete antigens that are similar to myelin self-antigens. This process, called molecular mimicry, provides a reason for the initiation of the autoimmune cascade. Furthermore, an alternative hypothesis may be that naturally present T-cells, which are specific for myelin antigens, begin to grown in numbers, which eventually makes them pathogenic. However, as mentioned earlier, these pathways, including environmental and genetic factors, seem to correlate more with a role in driving the autoimmune cascade, rather than initiating it.

Viewing these theories through a holistic perspective, one main factor that remains unexplained is the initial step or insult in the onset of MS. The previous paragraphs have indicated pathways in which immune cells are recruited into the brain in response to antigens, but the question we should be asking is how is the antigen formed in the first place? Is the CP important in the transfer of this antigen to the periphery? And if so, can treatments similar to cancer immunotherapy serve as a basis for cures? It has been postulated that MS may begin through infections or may have a genetic background. Interestingly, one study indicates another possible mechanism that may be responsible for the onset of MS.[32] Although this study presents information on the lacrimal gland epithelial cells and how they are affected by the autoimmune disease Sjögrens syndrome, this may be relevant to cells in the brain, as both locations are immune-privileged sites and contain their own inherent immune system. Moreover, the lacrimal epithelial cells as well as the CP epithelial cells play an important role in preserving the eye and the brain, respectively. This is done by mediating transfer of nutrients, secreting proteins that protect and regenerate damaged tissue. Thus, because of the relative similarity between both parenchymas, their respective epithelial layers, and the autoimmune nature of their respective disorders, it would be plausible to attribute conclusions from the eye to that of the brain.

In this study, lacrimal gland epithelial cells were constantly stimulated with carbachol, an agonist of acetylcholine. This was done in order to create an *in vitro* model of what occurs in patients with

Sjögrens syndrome, in which the lacrimal gland epithelial cells are constantly innervated by acetylcholine. It was found that this constant stimulation significantly altered the secretion pathways and trafficking of proteins in the lacrimal endothelial cells. It was observed that due to constant stimulation of the cell, traffic to the late endosome was blocked, thus resulting in accumulation of products in the early endosome. In healthy cells, there is a normal trafficking pathway of molecules that will be secreted, such as self-antigens, as well as endogenous molecules that follow the subsequent pathway: from the endoplasmic reticulum (where the proteins are synthesized), to the trans-golgi network, to the recycling endosome, and finally to the early endosome. It is here where deviations occur, with the initiation of multivesicular body formation for secreted particles and transfer of hydrolases and proteases to the late endosome and finally to the lysosome where it will be used to degraded organelles and macromolecules by products of the cell.

Moreover, there has been evidence that trafficking of lysosomal proteins can also proceed directly from the trans-golgi network to the late endosome. In a cell that is constantly stimulated, normal trafficking pathways are disrupted and transport of vesicles from the early endosome to the late endosome is halted. Thus, the proteins, including self-antigens that are to be secreted, and the proteases that should be transported to the lysosome, accumulate in the early endosome. Furthermore, as the late endosome forms the last vesicles with the remaining proteins inside of it, it begins to disappear as its size is not maintained due to the halt of traffic of new vesicles that would replenish the membrane of the endosome. The accumulation of these particles results in structural changes in the secreted auto-antigens, as they are preprocessed in the early endosome by the active proteases. These preprocessed antigens are then secreted into the extracellular matrix where they are they taken up by dendritic cells. Inside the dendritic cells, further processing of the antigen is done and eventually they are bound to MHC class II receptors and displayed on the surface where they may be presented to leukocytes.

In a normal situation, antigens that are released from cells have a distinct, or dominant, epitope, that when processed by dendritic cells, is always displayed on the MHC receptor. However, in a diseased situation, a preprocessed antigen is further processed in the dendritic cell, and thus, an incorrect epitope is displayed on its surface. Thus, this incorrect epitope is then presented to the leukocyte, which identifies this antigen as foreign and therefore becomes activated and cytotoxic. The cell becomes autoreactive and targets the lacrimal epithelial cell, initiating the autoimmune cascade. This mechanism, specifically the formation of a preprocessed antigen, may serve as the pathway for the initiation of autoimmune diseases. Although this is present in the lacrimal gland, it may shed light on a similar, if not identical, mechanism in the brain, which leads to MS because of the similarity in the immunological processes in the eye and the brain.[32]

With this, we have identified a potential mechanism that may explain the initial insult in the onset of MS. However, how does this translate to the brain? Furthermore, how is the choroid plexus involved in this pathway?

Chronic stimulation of lacrimal glands causes dysfunction of the trafficking pathways and subsequent release of preprocessed antigens, which may lead to the initiation of the autoimmune disorder, Sjögrens syndrome. In the brain, using the data and observations from the paragraphs earlier, we can hypothesize certain pathways and interactions that may initiate multiple sclerosis and how the choroid plexus may be involved. As noted previously, a current hypothesis is that chronic inflammation in the periphery initiates transcriptional changes in the choroid plexus.[21] These changes cause the upregulation of specific markers that control infiltration of leukocytes into the brain, an extremely important part of MS pathogenesis. Using this observation as the basis of our hypothesis of MS formation and judging by the age group (20–50 years of age, with very rare cases of pediatric MS) that is affected by MS, one can attribute stress and its hormone, cortisol, as one of the primary instigators of chronic inflammation in the human body. The population inside of this age group tends to have more stress in their lives due to school and work. High amounts of stress initiate the release of cortisol from the adrenal cortex. A stable amount of cortisol is known to activate antistress and anti-inflammatory pathways, and to initiate the upregulation of T_h2 cells, while downregulating the inflammatory T_h1 cells.[33] This shows that in tolerable amounts, cortisol can have beneficial effects on the immune system without affecting the body, especially the neural parenchyma. As mentioned previously, acute inflammation in the periphery causes genetic changes in the choroid plexus, but quickly returns to normal in a few days. However, in chronic stress situations, when the concentration of cortisol dramatically increases, many dire side effects can occur. Studies have shown that cortisol can cause inflammation in the periphery, and can cause overstimulation of neurons in the brain.[34,35] As cortisol travels to the brain and binds to receptors on neurons, this causes calcium to be released, which raises the basal membrane voltage of the cell and forces them to fire more often. This overstimulation eventually kills the cell. Studies have shown that this is most prevalent in the hippocampus, an important brain region that is proximal to the CP.[36] Moreover, cortisol further inhibits the rate of neurogenesis, which occurs in the SVZ and SGZ near the hippocampus.[37,38] With this, we can already see how the choroid plexus may be involved in these issues due to its proximity to the dysfunction. Furthermore, as stated previously, chronic inflammation in the body can lead to expression of markers that can increase the infiltration of leukocytes into the brain. Although stress is the initiator of overstimulation in the brain, other disorders can further increase stimulation of neurons. For example, epilepsy, characterized by chronic seizures, occurs due to over-reactive firing of the neuronal infrastructure. Interestingly, studies have shown that patients who experience seizures or who have epilepsy are more prone to be diagnosed with MS compared to the normal population.[39] This sheds light on the role of overstimulation of the brain in the pathogenesis of MS.

We highlighted two different pathways that could cause overstimulation of the neurons in the brain. This follows suit with the observation that overstimulation of the lacrimal gland may cause the onset of Sjögrens syndrome in the eye. Now that we have identified possible mechanisms of overstimulation in the brain, we can now continue to hypothesize the subsequent effects and the role of the CP. As overstimulation begins to affect the neurons, this will seemingly affect the oligodendrocytes, which are closely associated with these cells. What might be occurring may be similar to what was explained earlier in the onset of Sjögrens syndrome. Overstimulation of the cells causes release of preprocessed antigens by oligodendrocytes, which will be taken up by circulating dendritic cells or by the epiplexus cells in the CP and displayed on MHC class II receptors. Simultaneously, the consistent cell death associated with stress or overstimulation will also release factors that will be recognized by the resident T-cells in the CP. This will initiate the release of neurotrophic factors to help promote neurogenesis. However, as mentioned previously, neurogenesis will be inhibited by the presence of cortisol. Thus, these neurotrophic factors will not be ingested by the cells and they will accumulate in the CSF. Studies have shown that when inhibition of neurogenesis and blocking of neurotrophic factors occur, T-cells in the CP shift to a proinflammatory state. In conjunction with this, the preprocessed antigens presented to dormant T-cells in the epithelium will be characterized as pathogenic and will subsequently active them. To propagate the cycle, and continue to drive it, cortisol, as mentioned earlier, stimulates the T_h2 subtype, which will therefore recruit more cytotoxic T-cells from the periphery. Moreover, the entry of these recruited lymphocytes into the neural parenchyma will be more efficient because of the fact that chronic inflammation from the periphery, through the cortisol mechanism, had already induced genetic changes in the CP, which had upregulated the expression of infiltration markers.

The mechanism just presented depicts a possible scenario and a novel perspective in the onset of multiple sclerosis through cortisol and its subsequent overstimulation and modification of the CP. Not only does this mechanism portray the importance of the CP in MS, but it also takes into account the ease at which the CP can be infiltrated. The study mentioned previously explained that demyelination of the DGM occurs in early stages of MS and sometimes precedes the formation of white matter lesions, which may inherently indicate the presence of cytotoxic leukocytes in the ventricles and CSF and thus the formation of plaques in the DGM. The pathway suggested earlier fits this assertion because the CP can be easily manipulated to allow infiltration of peripheral immune cells. While the BBB can also allow this to occur, the process for the infiltration is more drawn out than that of the CP. For the BBB to become permeable, the astrocyte would initially have to pick up the antigen, secrete cytokines that would cause changes in the gene expression patters of BBB endothelial cells, which would then finally upregulate adhesion markers on the cell surface. Then the APC would need to go present the pathogenic antigen to a leukocyte, which would then initiate the immune cascade. On the other hand, the CP intrinsically contains all the molecular necessities that the onset of the disease can exploit to expedite its process. First of all, antigens would be able to be detected much faster in the CSF rather than being detected by random astrocyte migration. Because of the presence of dormant T-cells that reside in the epithelial cells of the CP, there would be no need to extend to the periphery to communicate with immune cells. The antigens would immediately be present to the naïve T-cells and cause subsequent differentiation into Th1 and

Th2 subtypes. Th1 would secrete IFN-γ and induce upregulation of adhesion markers and Th2 would recruit mature immune cells from the periphery. Other subtypes of the differentiated T-cell, such as Th17, would simultaneously enter the neural parenchyma and begin the autoimmune cascade. Thus, because of the presence of all the molecular machinery in the CP and the lack thereof in the BBB, this provides a solid rationale of the importance of the CP in the initiation of MS.

FUTURE DIRECTIONS

We have so far focused on identifying a possible model of pathogenesis and subsequent autoimmune reactions. We now turn our focus on using the CP as a possible route of entry for immunotherapy. Seeing how the disease can exploit the resources in the CP, a strategy would be for us to think along the same lines, to exploit the CP, in order to find methods to enhance the treatment of MS and possibly design more effective cures. Current treatments for MS are divided into two groups, one affecting the acute symptoms of the disease while the other group is based more on long-term treatments. The latter focuses on preventing progression of the disease and future relapses. During the initial stages of drug development for MS, compounds that focused on interfering or suppressing the immune response were favorable. Recently, developments in drugs have created more effective treatments. However, these advancements have high risk of safety issues.

In light of these developments, it may be more plausible to focus on a more holistic treatment rather than focusing on suppressing certain molecules and genes. One possible treatment utilizing immunotherapy would be to specifically select for lesion-resistant oligodendrocytes that are derived from the patient. For example, a patient with MS will have oligodendrocytes removed from their neural parenchyma. These cells will then be cocultured with leukocytes that have also been removed from the brain. These cells can come from either the choroid plexus, which we have shown to be in high concentrations during neural trauma, or by filtering the CSF, again another highly concentrated area, and selecting for these cells. Once the leukocytes and the oligodendrocytes have been isolated from the patient, these will then be cocultured to preselect for cells that are functionally sound and resistant to destruction by leukocytes. The mechanism behind this may lie in the fact that these oligodendrocytes do not express certain markers, which separate them from their damaged counterparts. Once these cells have been selected, they will be separated from the leukocytes and injected into the lesion sites of the patient. To ensure these cells are not destroyed, the patient will be immunosuppressed and will be slowly weaned off the medication until the immune system is fully functional. These newly implanted cells will thus hopefully be resistant to further leukocyte damage and will thus aid in remyelinating damaged neurons.

Before this possible treatment enters clinical trials, it would be important to have a subset of preclinical experiments that would check for specific markers at the lesion site. This would provide more information on the lesion microenvironment as well as on the physiological state of the present leukocytes, APCs, and oligodendrocytes. By discovering what markers are expressed on the oligodendrocytes, we can genetically engineer leukocytes that will express these antigens on their MHC class II receptor. Therefore, they would not be cytotoxic against the newly transplanted oligodendrocytes, thus increasing the efficacy of the treatment. Along with the immunosuppression of the innate immune system of the patient, this substitution of anticytotoxic leukocytes would in theory stop the degradation and destruction of neural tissue. With this, we would take advantage of the inherent behavior of leukocytes to be present in the CP, and thus would use the CP as a means of a stable structure in which the newly transplanted leukocytes can find solace. The oligodendrocytes would come to the aid of the dying neurons, and with these transplantations, we would have hopefully replaced a damaged function, with a new and MS-resistant immune system.

CONCLUSIONS

In this chapter, we have discussed the importance of the immune system in the neural parenchyma and highlighted a mechanism of infiltration into the brain by these cells through the choroid plexus. We also focused on discussing how the relative ease of infiltration may factor into the propagation of the pathogenesis in multiple sclerosis. Moreover, we presented a novel idea on the initial insult that causes the onset of MS and how this pathway is mediated by the choroid plexus. It is important to value the statements as mere speculation rather than concrete facts and thus utilize them to spur more ideas and experiments. In conclusion, the choroid plexus presents a fascinating microenvironment of the brain, which is grossly understudied, and may hold the key for understanding autoimmune neurodegenerative disorders.

References

1. Dantzer R, O'Connor JC, Freund GG, Johnson RW, Kelley KW. From inflammation to sickness and depression: when the immune system subjugates the brain. *Nat Rev Neurosci.* 2008;9:46–56.
2. Monje ML, Toda H, Palmer TD. Inflammatory blockade restores adult hippocampal neurogenesis. *Science.* 2003;302:1760–1765.
3. Raison CL, Capuron L, Miller AH. Cytokines sing the blues: inflammation and the pathogenesis of depression. *Trends Immunol.* 2006;27:24–31.
4. Ransohoff RM, Brown MA. Innate immunity in the central nervous system. *J Clin Invest.* 2012;1164–1171:122.4, http://www.ncbi.nlm.nih.gov/pubmed/22466658.
5. Ghadimi D, Fölster-Holst R, de Vrese M, et al. Effects of probiotic bacteria and their genomic DNA on TH1/TH2-cytokine production by peripheral blood mononuclear cells (PBMCs) of healthy and allergic subjects. *Immunobiology.* 2008;213:677–692.
6. Shechter R, Miller O, Yovel G, et al. Recruitment of beneficial M2 macrophages to injured spinal cord is orchestrated by remote brain choroid plexus. *Immunity.* 2013;555–569:38.3.
7. Romagnani S. T-cell subsets (Th1 versus Th2). *Ann Allergy Asthma Immunol.* 2000;9–21:85.1.
8. Baruch K, Schwartz M. CNS-specific T cells shape brain function via the choroid plexus. *Brain Behav Immun.* 2013;34:11–16:http://www.sciencedirect.com/science/article/pii/S0889159113001426.
9. Peterson DA, DiPaolo RJ, Kanagawa O, Unanue ER. Cutting edge: negative selection of immature thymocytes by a few peptide–MHC complexes. Differential sensitivity of immature and mature T cells. *J Immunol.* 1999;162:3117–3120.
10. Kipnis J, Cohen H, Cardon M, Ziv Y, Schwartz M. T cell deficiency leads to cognitive dysfunction: implications for therapeutic vaccination for schizophrenia and other psychiatric conditions. *Proc Natl Acad Sci USA.* 2004;101:8180–8185.
11. Hauben E, Agranov E, Gothilf A, et al. Posttraumatic therapeutic vaccination with modified myelin self-antigen prevents complete paralysis while avoiding autoimmune disease. *J Clin Invest.* 2001;108:591–599.
12. Kipnis J, Gadani S, Derecki NC. Pro-cognitive properties of T cells. *Nat Rev Immunol.* 2012;12:663–669.
13. Schwartz M, Shechter R. Protective autoimmunity functions by intracranial immunosurveillance to support the mind: the missing link between health and disease. *Mol Psychiatr.* 2010;15:342–354.
14. Johanson CE, Stopa EG, McMillan PN. The blood–cerebrospinal fluid barrier: structure and functional significance. *Methods Mol Biol.* 2011;686:101–131.

15. Szmydynger-Chodobska J, Strazielle N, Gandy JR, et al. Posttraumatic invasion of monocytes across the blood-cerebrospinal fluid barrier. *J Cereb Blood Flow Metab.* 2012;32:93–104.

16. Engelhardt B, Ransohoff RM. The ins and outs of T-lymphocyte trafficking to the CNS: anatomical sites and molecular mechanisms. *Trends Immunol.* 2005;26:485–495.

17. Ching S, He L, Lai W, Quan N. IL-1 Type I receptor plays a key role in mediating the recruitment of leukocytes into the central nervous system. *Brain Behav Immun.* 2005;127–137:19.2.

18. Chodobski A, Szmydynger-Chodobska J. Choroid plexus: target for polypeptides and site of their synthesis. *Microsc Res Tech.* 2001;52:65–82.

19. Sathyanesan M, Girgenti MJ, Banasr M, et al. A molecular characterization of the choroid plexus and stress-induced gene regulation. *Transl Psychiatr.* 2012;2:e139.

20. Marques F, Falcao AM, Sousa JC, et al. Altered iron metabolism is part of the choroid plexus response to peripheral inflammation. *Endocrinology.* 2009;150:2822–2828.

21. Marques F, Sousa JC, Coppola G, et al. The choroid plexus response to a repeated peripheral inflammatory stimulus. *BMC Neurosci.* 2009;10:135.

22. Compston A, Coles A. Multiple sclerosis. *Lancet.* 2002;359:1221–1231.

23. Murray ED, Buttner EA, Price BH. Depression and psychosis in neurological practice. In: Daroff R, Fenichel G, Jankovic J, Mazziotta J, eds. *Bradley's Neurology in Clinical Practice.* Philadelphia, PA: Elsevier/Saunders; 2012.

24. Dyment DA, Ebers GC, Sadovnick AD. Genetics of multiple sclerosis. *Lancet Neurol.* 2004;3:104–110.

25. Vercellino M, Masera S, Lorenzatti S, et al. Demyelination, inflammation, and neurodegeneration in multiple sclerosis deep gray matter. *J Neuropathol Exp Neurol.* 2009;489–502:68.5.

26. Brown DA, Sawchenko PE. Time course and distribution of inflammatory and neurodegenerative events suggest structural bases for the pathogenesis of experimental autoimmune encephalomyelitis. *J Comp Neurol.* 2007;502:236–260.

27. Lowenstein PR. Immunology of viral-vector-mediated gene transfer into the brain: an evolutionary and developmental perspective. *Trends Immunol.* 2002;23:23–30.

28. Vercellino M, Votta B, Condello C, et al. Involvement of the choroid plexus in multiple sclerosis autoimmune inflammation: a neuropathological study. *J Neuroimmunol.* 2008;133–141:199.1–2.

29. Engelhardt B, Ransohoff RM. Capture, crawl, cross: the T cell code to breach the blood–brain barriers. *Trends Immunol.* 2012;579–589:33.12, http://www.ncbi.nlm.nih.gov/pubmed/22926201.

30. Pedemonte E, Mancardi G, Giunti D, et al. Mechanisms of the adaptive immune response inside the central nervous system during inflammatory and autoimmune diseases. *Pharmacol Ther.* 2006;111:555–566.

31. Kivisäkk P, Imitola J, Rasmussen S, et al. Localizing central nervous system immune surveillance: meningeal antigen-presenting cells activate T cells during experimental autoimmune encephalomyelitis. *Ann Neurol.* 2009;457–469:65.4.

32. Rose CM, Qian L, Hakim L, et al. Accumulation of catalytically active proteases in lacrimal gland acinar cell endosomes during chronic ex vivo muscarinic receptor stimulation. *Scand J Immunol*. 2005;36–50:61.1.

33. Elenkov IJ. Glucocorticoids and the Th1/Th2 balance. *Ann NY Acad Sci*. 2004;1024:138–146.

34. Frodl T, O'Keane V. How does the brain deal with cumulative stress? A review with focus on developmental stress, HPA axis function and hippocampal structure in humans. *Neurobiol Dis*. 2013;52:24–37.

35. Maric NP, Adzic M. Pharmacological modulation of HPA axis in depression – new avenues for potential therapeutic benefits. *Psychiatr Danub*. 2013;25: 299–305.

36. Lee AL, Ogle WO, Sapolsky RM. Stress and depression: possible links to neuron death in the hippocampus. *Bipolar Disord*. 2002;4:117–128.

37. Leonard B, Maes M. Mechanistic explanations how cell-mediated immune activation, inflammation and oxidative and nitrosative stress pathways and their sequels and concomitants play a role in the pathophysiology of unipolar depression. *Neurosci Biobehav Rev*. 2012;36:764–785.

38. Maes M, Yirmyia R, Noraberg J, et al. The inflammatory & neurodegenerative (I&ND) hypothesis of depression: leads for future research and new drug developments in depression. *Metab Brain Dis*. 2009;24:27–53.

39. Allen AN, Seminog OO, Goldacre MJ. Association between multiple sclerosis and epilepsy: large population-based record-linkage studies. *BMC Neurol*. 2013;189:13.1.

CHAPTER

8

The Choroid Plexus and Cerebrospinal Fluid System: Roles in Neurodegenerative Diseases

Jérôme Badaut*, Jean-François Ghersi-Egea[†]

*UMR 5287-Institut de Neurosciences Cognitives et Intégratives d'Aquitaine, Université de Bordeaux, Bordeaux, France; [†]Blood–Brain Interface Group, Oncoflam Team, and BIP Platform, INSERM U1028, CNRS UMR5292, Lyon Neuroscience Research Center, Faculté de Médecine RTH Laennec, Lyon, France

OUTLINE

The Choroid Plexus and Cerebrospinal Fluid. http://dx.doi.org/10.1016/B978-0-12-801740-1.00008-1

129

INTRODUCTION

Research on the pathophysiology of neurodegenerative diseases has been very intense in the last few decades with an overwhelming neurocentric focus on molecular mechanisms. This effort has not resulted in effective therapeutics capable of modifying disease outcomes, in particular for Alzheimer's disease (AD). Recently, non-neuronal cells, such as astrocytes, endothelial cells, and choroid plexus (CP) epithelial cells, have been emerging as potentially important players in the pathophysiology of neurodegenerative processes.[1,2] The chapter is focused on the key functions of CP, cerebrospinal fluid (CSF), and interstitial fluid (ISF) filling the extracellular space (ECS), and their implications in neurodegenerative courses.

PART 1: CHOROID PLEXUS, CEREBROSPINAL FLUID, INTERSTITIAL FLUID: RELATIONSHIP WITH GLYMPHATIC SYSTEM

Choroid Plexus Anatomy, Transport, and Secretion Functions

The fine control of cerebral extracellular fluid homeostasis necessary for central nervous system (CNS) function requires a strict regulation of the molecular and cellular exchanges between the brain and body. This regulation results from the specific properties of blood–brain interfaces, that is, the endothelium of the cerebral microvessels and the epithelium of the CPs, in combination with the CSF circulatory system. The latter is unique to the brain, which is otherwise devoid of lymphatic drainage.[3] There are four CPs located in the ventricular system of the brain. They form the interface between the blood and CSF that they secrete. Two CPs originate from the choroidal fissure, and lie in the central and temporal horn of the two lateral ventricles. They join the CP appended to the ceiling of the third ventricle (and pineal recess in some species including rodents) through the foramen of Monro. The fourth CP is a distinct entity located in the fourth ventricle. It extends centrally through the foramen of Magendie, and laterally into the lateral recesses of the fourth ventricle, reaching the subarachnoid spaces through the foramen of Luschka. Macroscopically, the different CPs display a different morphology. Those of the lateral ventricle form a veil-like structure while the fourth ventricle CP is lobulated and bulky.

The embryonic origin of the CPs is dual. They originate from neuroectodermic cells and underlying mesodermic cells of the neural tube and appear early during development (from day 45 of gestation in humans). Rapidly, they form multiple villi organized as ramified papillary structures, which after four phases of development ultimately organize as a simple epithelium of ectodermic origin surrounding a highly vascularized conjunctive stroma of leptomeningeal origin.[4] These villi present the following histological and ultrastructural characteristics (Fig. 8.1). The choroidal epithelial cells, in direct contact with the ventricular CSF by their apical pole, are 10-μm thick, lie on a continuous basal membrane, and present numerous apical cytoplasmic projections or microvilli (brush border), a few cilia, and a complex interdigitated basolateral membrane network. These characteristics extend the surface area available for exchange between the CSF and the choroidal stroma strongly. The total apical surface is estimated to be 75 cm^2 in a one-month-old rat, a figure close to the surface area developed by the blood–brain barrier in the same animal.[5] In humans, the same choroidal versus endothelial surface area ratio would be close to 1:10. The epithelial cells present numerous mitochondria, a developed endoplasmic reticulum, and numerous vesicles, all indices of their intense metabolic and secretory activities. The choroidal stroma is richly vascularized. In humans the blood is supplied to the lateral ventricle CPs by the anterior and posterior choroidal arteries, and drained by the corresponding choroidal veins. The posterior artery also irrigates the third ventricle CP, while the fourth ventricle tissue is perfused through the posterior–inferior cerebellar artery. The penetrating choroidal larger

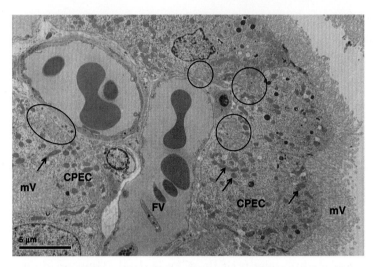

FIGURE 8.1 **Histological features of the blood CSF barrier.** An Electron micrograph of a section through a villus of the rat fourth ventricle choroid plexus is shown. The choroid plexus epithelial cells (CPEC), display a large surface area of exchange with the CSF and the blood contained in the fenestrated vessels (FV), owing to their apical brush border formed by microvilli (mV) and to the interdigitations of their basolateral membrane (circles). Arrows points to the numerous mitochondria highlighting the strong metabolic capacity of the cells.

vessels keep a barrier-type phenotype, while the terminal vascular loops that irrigate each villi are dilated capillaries, with a thin endothelial border that lie on a continuous basal membrane and present numerous fenestrations. This fenestrated phenotype results from the continuous exposure of endothelial cells to vascular endothelial growth factor (VEGF) secreted by the adjacent choroidal epithelium. The blood flow, measured in different species, is higher in the choroidal tissue than in any other region of the brain. In the rat it reaches 3–5 mL/g·min, that is, three to five times higher than the cortical blood flow.[6] The exact meaning of this high blood flow is not well understood. In particular blood flow is not a limiting factor for CSF secretion.[7] Finally, immune cells of the lymphoid and myeloid lineage are also associated with the CPs. In rats as in humans, dendritic cells, macrophages, and T lymphocytes are found in the choroidal stroma.[8–11] Some dendritic cells present an intraepithelial localization.[12] Myeloid progenitors able to differentiate in macrophages and predendritic cells were identified in the stroma.[13] This suggests that the CP stroma represents a reservoir for macrophagic cells. Phagocytic cells called Kolmer or epiplexus cells have also been consistently reported as crawling on top of the microvilli at the ventricular surface of the CP epithelium among species.[14,15] The precise identity and function of these peculiar cells, which also display dendritic markers, remains elusive.[9]

Owing to the presence of tight junctions that link the choroidal epithelial cells together, and to the highly permissive fenestrated stromal vessels, the choroidal epithelium layer forms the actual barrier between the blood and the CSF. This barrier is not as tight as the blood–brain barrier. Its electrical resistance is of a few tenths/hundreds ohms, while the BBB electric resistance reaches thousands of ohms. *In vivo* experiments have shown that the blood-CSF barrier (BCSFB) also allows a low but detectable permeation of polar tracers.[16] This likely results from the peculiar molecular composition of the junctional protein complexes, which in addition to the classical occludin and JAM (junctional adhesion molecule), include several claudins not found at the BBB, among them claudin-2, a monovalent cation selective pore-forming junctional protein that may also act as a water channel (reviewed in Ref. 17). Nonetheless, the choroidal tight junctions strongly impede paracellular transfer of polar compounds between the blood and the CSF. The junctional complexes also participate to the establishment of cell polarity, which is key to the transport and secretion functions of the CP, in particular CSF secretion. Another difference between BCSFB and BBB is the presence of the astrocyte end-feet layer that unsheathes the cerebral blood vessels. Besides tight junctions, the presence of the extracellular matrix (ECM) and the layer of astrocyte end-feet is enough to create an additional physical barrier between the endothelium and the neuropil, and for example to maintain antigen-presenting myeloid cells in perivascular space in physiological states.[18]

The rate of CSF secretion by the CP is relatively high, 0.2–0.3 $mL \cdot g^{-1} \cdot min^{-1}$ in humans. This secretion results from an active and regulated process that guarantees a relatively stable, slightly hyperosmotic composition of the nascent CSF. The current model of CSF secretion is described by Damkier et al.[19] The driving force is a unidirectional active transcellular blood-to-CSF flux of Na^+, Cl^-, and HCO_3^-, which creates an osmotic gradient followed by water movement. This active ionic flux is created by carbonic anhydrase and Na^+/K^+-ATPase. The latter enzyme presents a peculiar apical localization, a feature not found in other epithelia. A large number of channels and transporters for monovalent inorganic ions participate in the secretory process. The model also implies that a transfer of K^+, necessary to provide adequate K^+ levels in the CSF, occurs via the paracellular route. Water movement likely involves aquaporin 1 and possibly 4 (see later text for more details on aquaporin 4), both of these channels have been found on epithelial cells of CP (Fig. 8.2).[20,21] AQP1 has been proposed to be a major channel involved in CSF formation.[21] Water movement may also occur through the paracellular pathway. The role of pore-forming claudins at the BCSFB remains to be evaluated in this context of CSF secretion.[17]

In addition to the secretion of CSF and control of inorganic ion balance in brain fluids, the CPs participate in the maintenance of cerebral homeostasis by acting as a protective barrier toward potentially deleterious organic compounds. Tight junctions strongly impede nonspecific paracellular leakage from blood to the CSF. A number of transporters of broad specificity act as efflux carriers at the blood–CSF barrier. They prevent the transcellular entry of lipophilic blood-borne chemicals, or accelerate the efflux of CSF-borne chemicals and endogenous metabolites. These transporters either possess an ATP binding site and belong to the ATP-binding cassette (ABC) protein superfamily, or are solute transporters, mainly of the Slco and Slc22 families. The most documented efflux transporters are Abcc1, Abcc4, Slco1a5, and Slc22a8. They have different but overlapping specificities, and accept many structurally unrelated environmental toxic compounds or drugs, including anti-inflammatory, antibiotics and antiviral drugs, antineoplastic drugs, some antiepileptics, antidepressant and psychotropic agents, and drug conjugates. They also can carry steroids and lipid immunomodulators, thus participating in the control of the cerebral levels of these biologically active mediators (reviewed by Strazielle et al.[22]). They are differentially localized on the two membranes, as a function of their directionality of transport (summarized in Fig. 8.3). Immunoglobulin transfer likely involving the Fc neonatal receptor occurs across the blood–CSF barrier. It is unidirectional, in the CSF-to-blood direction. This efflux mechanism therefore confers a role for the blood–CSF barrier in the control of immunoglobulin levels into the brain.[23] The CPs also express a number of drug metabolizing and antioxidant enzymes. These enzymes inactivate reactive molecules *en passage* through

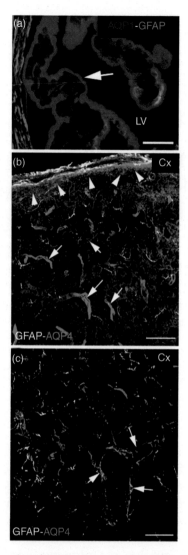

FIGURE 8.2   **AQP1 and 4 distributions in the CNS.** Imunolabeling pictures for AQP1 (a) and AQP4 (b, c) present in the choroid plexus (a) and in the cortex (b, c) (adapted from Ref. [20]). (a) AQP1 immunostaining (red) in choroid plexus epithelium (arrow) located in the lateral ventricle (LV). The border of the LV is outlined by the GFAP staining (green), specific marker of the astrocytes. (b) AQP4 labeling (green) in the parietal cortex (Cx) is abundant on the glia limitans (arrow heads), revealed in gray by the GFAP staining. The AQP4 labeling is underlining the blood vessels (arrows) by its concentration in the astrocyte end-feet stained by the anti-GFAP (gray). (c) AQP4 distribution in the deeper cortex layer showed the "polarization" of the AQP4 labeling around the blood vessels (arrows). The double staining AQP4 and GFAP exhibits its presence on the astrocyte endfeet (arrows). Bar a, b, c = 50 μm.

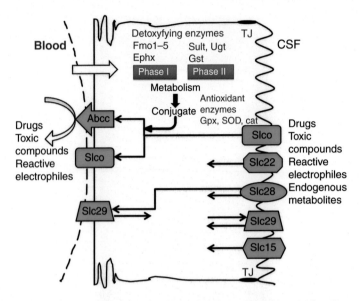

FIGURE 8.3 Schematic representation of neuroprotective mechanisms at the blood–CSF barrier. In addition to tight junctions (red) that prevent nonspecific paracellular diffusion, a wide variety of multispecific efflux transporters (green), as well as antioxidant and drug metabolizing enzymes (blue), participate in the neuroprotective and detoxifying functions of the choroid plexus. Only the most documented mechanisms are represented on the scheme. *Reprinted with permission from.*[26]

the blood–CSF barrier. Inactive metabolites that are generated can be extruded by efflux transporters into the blood (Fig. 8.3).[24,25]

Besides neuroprotection, the blood–CSF barrier fulfills an influx transport function for selected nutrients (e.g., glucose and large neutral amino acids), divalent inorganic anions (e.g., calcium and manganese), and micronutrients (e.g., vitamin C, folic acid, and selected nucleosides).[27,28] Polypeptide and protein transfer across the blood–CSF barrier also occurs and uses different types of mechanisms. It can involve receptor-mediated transcytosis. For instance, the leptin receptor, transferrin receptor, insulin receptor, or else low-density lipoprotein (LDL) receptor-related proteins (LRPs) that accept several biologically active polypeptides as ligands in addition to LDL, display a robust expression at the CP. The ability of these different receptors to induce complete transcytosis rather than endocytosis remains, however, a subject of debate (reviewed in[23]). The transfer of plasma proteins into the CSF has also been described across selected cells of the choroidal epithelium. The exact pathway remains to be clearly defined. It is predominant in the developing brain whose CSF plasma-protein concentration is higher than in the adult, but remains active in adult animal (reviewed in[23]). Finally, a large number of polypeptides are directly synthesized and secreted by the CP epithelial cells.[29] They include the thyroid hormone transport protein transthyretin, various growth factors (e.g., insulin-like growth factor II, prostaglandin D synthase, and Klotho protein), and guidance molecules (e.g., Slit-2, the protease inhibitor cystatin C, gelsolin, and vasopressin). Most of them are mainly secreted at the apical membrane, and thus into the CSF, with the exception of VEGF.

Cerebrospinal Fluid Circulation

Polypeptides synthesized by, and molecules transported through, the CPs can be delivered at various sites of the brain through the CSF circulation. Previously seen as a simple passive drainage system for the brain, the CSF circulation is increasingly recognized as participating in volume transmission throughout the brain, and may also be an important actor of clearance of brain metabolites from parenchyma through its connection with the perivascular fluid circulation (shown later). The recent revision on the CSF circulation with its dynamic introduces the idea that the CSF and CP dysfunctions could be part of the pathophysiological process in neurodegenerative diseases.[30]

In humans as in other mammals, the CSF represents half of the extracellular fluid of the brain. In humans, the volume of CSF is estimated to be 150–270 mL, and is renewed several times a day. The active secretion of CSF by the CPs is the driving force for CSF flow. The ciliated ependyma also contribute to the flow.[31] The CSF produced by the CP lying in the different horns of the lateral ventricles, the pineal recess, and third ventricle circulates through the anterior and posterior parts of the third ventricle into the aqueduct, to reach the small mesencephalic ventricle abutting the cerebellum and the fourth ventricle. CSF circulating through the ventral part of the third ventricle located within the hypothalamus also feeds the optic, infundibular, and mamilliary recesses, before reaching the aqueduct of Sylvius. From the fourth ventricle, the fluid flow, increased by CSF produced locally by the CP, is directed toward the central canal of the spinal cord, the cisterna magna (through the foramen of Magendie), and the lateral recesses of the fourth ventricle. It then reaches the caudal subarachnoid spaces of the brain through the foramens of Luschka. Tracer experiments have shown that CSF flow is brisk along the ventricular system, and that the fluid transits from the lateral ventricles to the cisterna magna within minutes. It slows while traveling through the subarachnoid and cisternal spaces, mainly through the ambient and quadrigeminal cisterns. These cisterns are fluid-filled internal spaces located between the midbrain and hippocampi/cortices and contain trabeculae of meningeal origin, veins, and arteries.[32] CSF also circulates from the fourth ventricle along the spinal cord subarachnoid spaces, and ventrally in a caudo-rostral direction through the cerebelopontine, interpeduncular, optic tract, and laminae terminalis cisterns, and around the olfactory bulbs. In rodents, very little, if any, CSF circulates at the surface of the cerebral cortex, probably as a result of the paucity of cortical sulci. An elegant study recently allowed visualizing the subarachnoid/cisternal pathways of convective flow *in vivo* by MRI in the rat.[33] Some confusion may arise, however, from the terms used by the authors who refer to this pathway as a "paravascular route," and refer to the optic tract and quadrigeminal cisterns as pituitary recess and pineal gland recess, respectively, while these names usually correspond to ventricular extensions. For illustration of these different fluid-filled cerebral spaces, see Schmitt et al. and Ghersi-Egea et al.[10,32] Of note, the third and the mesencephalic ventricles are separated from the velum interpositum and the superior medullary velum, respectively, only by a thin layer of cells. A convective flow between these compartments allows material borne by ventricular CSF to reach directly these leptomeningeal internal spaces, and from there the ambient and quadrigeminal cisterns.[32] Ultimately, the CSF is absorbed in the venous blood at the arachnoid villi or drained along the cranial nerves into the lymphatic system. The proportion of these two routes of resorption is not known and likely differs among species.

Exchanges between the ventricular CSF and the brain parenchyma occur across the more permissive ependyma. In most places ependyma is formed by a single layer of cuboidal cells linked by adherens, but not tight junctions, thus facilitating paracellular diffusion of lipid-insoluble molecules into the brain parenchyma. Intracellular retrograde transport of CSF-borne compounds into deeper structures also occurs, as exemplified for manganese supply to the brain.[34] The ependyma delimiting the circumventricular organs, specialized in sensing blood-borne hormones and other products, and lacking a blood–brain barrier, displays specific properties and in particular a tight phenotype.[35,36] The exchanges between the subarachnoid/cisternal CSF and the underlying neuropil are complex. They occur across the permeable pia matter and glia limitans lying immediately underneath. The glia limitans is formed by a complex network of glial cell membranes that slow down diffusional processes, but may at the same time act as a reservoir for active compounds.[32] Finally, the CSF mixes with the perivascular fluid of penetrating arteries and veins. Arteries, but not veins, traveling through the cisternal and subarachnoid spaces, are covered by a membrane of leptomeningeal origin, thus delimiting an intermediate space in which CSF-borne tracers, including peptides such as Aβ, remains trapped.[32,37] These spaces are probably connected with the Virchow–Robin spaces surrounding the arteries penetrating the neuropil. These specific sites of fluid flow and exchanges are described further in the next paragraph.

Paravascular Flow and Glymphatic System: Definition

Cerebrospinal fluid found in the subarachnoid space and ISF filling the ECS presents a similar composition. Brain ECS is not only filled by ICS but it is also composed by ECM, thought to be a mesh-like network organized around hyalorunic acid, heparin sulfate, proteoglycans, and proteins like laminins and collagens.[38] ECM components are bound to the cells and free floating in the ECS. This ECM mesh could regulate the diffusion of the ISF in the brain ECS by increasing the viscosity, by regulating the geometry of the ECS, by the repulsion, attraction, transient binding or sequestration of ions and other charged substrates.[38,39] It is generally accepted that bulk flow of fluid is occurring within the ECS. The ECS allow a diffusion of small compounds with an estimated space widths in the range of 40–60 nm in rodents. The perivascular route represents a lower resistance pathway with perivascular space widths of about 5–10 μm.[39] Anatomically, perivascular spaces, also called Virchow–Robin spaces, are surrounding penetrating vessels (artery and veins). Following the branching of vessels, Virchow–Robin spaces are decreased at the arterioles and venules. It is not excluded that a perivascular space is still present at the capillary level.[40–42] This compartment is filled with CSF that surrounds subarachnoid and penetrating blood vessels of the brain. At the capillary level, exchanges between CSF and ISF are possibly happening, which facilitates the clearance of interstitial solutes and wastes from the brain parenchyma.[40–43] Previously, Rennels et al. reported that subarachnoid CSF enters the brain along perivascular spaces surrounding cortical penetrating arteries, traveling along the abluminal side of vessels to reach the basal lamina of the terminal cerebral capillary bed.[44] Recently, the concept of CSF flow from the subarachnoid space along the vasculature was confirmed by *in vivo* two-photon microscopy,[45,46] by *ex vivo* fluorescence imaging,[47] and by dynamic contrast-enhanced magnetic resonance imaging.[33,48] A significant proportion of subarachnoid CSF rapidly re-enters the brain along perivascular pathways surrounding penetrating arteries, reaching the terminal capillary beds, and exchanging with ISF throughout the brain. It is important to underline that the direction of the CSF movement along the brain vasculature is still debated and the speed of the flux seems to be high due to the experimental design. Injection of tracers by pressure could force the flux in the CSF compartment. It is also important to note that it is likely that the CSF is going out along the perivascular space present along the veins. The movement of the CSF along the arteries is dependent on the arterial pulsatility.[49]

Interestingly, this mechanism of CSF and ISF exchange is associated with aquaporin-4 (AQP4), a water channel, present on all astrocyte end-feet embedding the entire vascular tree independently of the diameter of the blood vessel (Fig. 8.2).[20] This perivascular CSF–ISF exchange is facilitated by astroglial water transport via the AQP4 and supports the clearance of interstitial solutes, such as soluble Aβ, from the brain parenchyma. Based upon its similarity in function to the peripheral lymphatic system and its dependence on astroglial water flux, this network was termed the "glymphatic" system.[45] However, in regard to the role of the astrocytic AQP4, it is interesting to pinpoint that deletion of those channels has as a consequence a wider ECS in AQP4 knockout compared to wild-type mice.[50] Therefore, it looks like that the presence of the AQP4 is critical in the regulation of the size of the ECS, by facilitating the water movements.[20] In accordance with this observation, AQP4 expression increases during brain development in parallel with the decrease of the ECS size.[51] Interestingly, the glia limitans is the cellular structure at the interface between the CSF in subarachnoid space and brain parenchyma and expresses high level of AQP4.[20] It is also possible that AQP4 plays a role in exchanges between ISF and CSF at this interface.

In vivo and ex vivo analysis of fluorescent CSF tracer influx into and through the brain parenchyma demonstrate that reducing cerebral arterial pulsatility slowed paravascular CSF–ISF exchange in the brain, while increasing pulsatility with dobutamine accelerated the rate of perivascular CSF influx into brain tissue. These findings demonstrate that cerebral arterial pulsation is a significant driver of perivascular CSF–ISF exchange in the brain.[49] The dysfunction in the CSF–ISF exchange has been hypothesized to participate in the development of AD[45] and the hypothesis is supported with clinical observations showing that the CP–CSF system is altered in AD.[30] This aspect will be discussed more in details later in the chapter.

PART 2: CHOROID PLEXUS: INFLAMMATION AND NEURODEGENERATIVE DISEASES

The Choroid Plexus–CSF System in Neuroimmune Surveillance and in Primary Neuroinflammatory Diseases

The CP–CSF system appears to play a role, not only in the movement of solutes into and within the brain, but also in CNS trafficking of immune cells. The CSF is considered as the only site in the healthy brain that contains CD4+ T lymphocytes.[52] These cells are memory T cells and express high levels of the adhesion molecule P-selectin glycoprotein ligand 1 (PSGL-1).[53,54] P-selectin, a major counterligand for PSGL-1, is confined to the choroid plexus and the meningeal vessels as shown in mice and humans, and facilitates the early migration of T lymphocytes in the healthy mouse brain,[55] while the resting microvessel endothelium forming the blood–brain barrier does not support cell extravasation.[53,56] Leucocytes can in theory enter the CSF upstream via the choroid plexus into the ventricular space from where they follow the flow, or they can extravasate downstream, from subpial vessels into the subarachnoid spaces. Several observations support the former route through the choroid plexus, at least during normal immunosurveillance and in the early phase of neuroinflammatory processes. T cells are present in the human and murine choroid plexus stroma[54,57] and their numbers increased to some extent after nonspecific peripheral immune activation.[10,57] In addition, the initiation of experimental autoimmune encephalomyelitis, often used as an animal model of multiple sclerosis, requires brain entry of TH17 lymphocytes through the choroid plexus. Their penetration in the CNS is dependent on the chemokine receptor CCR6, whose chemokine ligand CCL20 is constitutively synthesized by the human and murine choroidal epithelium.[58] The relevance of this chemokine–chemokine receptor pair in transepithelial migration at the choroid plexus is underlined by the leukocyte accumulation within the conjunctive stroma of the CP in CCR6-deficient mice. Spatiotemporal analyses of the pathogenesis of murine and rodent experimental autoimmune encephalomyelitis indicate that periventricular structures are among the primary target areas of early T-cell and myeloid cell infiltration.[10,59] Homing of T cells into the CSF via the choroid plexus may similarly contribute to the preferential localization of focal demyelinated plaques in periventricular areas in patients with multiple sclerosis.[60] The exact mechanisms of cell recruitment across the choroid plexus, that is, extravasation across the choroidal vessels and cell trafficking from the stroma to the CSF space across the choroidal epithelium, remain to be elucidated.

Different primary neuroinflammatory and neuroinfectious diseases are characterized by elevated levels of CSF-borne chemokines (e.g., Ubogu et al.[61]). In addition to a potential action in amplifying T-cell traffic through the choroid plexus, chemokines circulating through the CSF may also contribute a local inflammatory environment around connected perivascular spaces of penetrating vessels, participating to the secondary immune cell invasion across the BBB-forming endothelium at these sites in neuroinflammatory disorders.[10] Taken together, the data indicates that the CP–CSF system participates in the orchestration of immune cell invasion into the CNS during neuroinflammation.

Myeloid cells associated with the choroidal tissue in physiologic conditions may also participate in neuroimmune surveillance as presenting cells for CSF-borne IgG, the latter transported by the receptor FcRn. This hypothesis has not yet been tested. Finally, in the context of viral infection, the choroidal myeloid cells appear to be a reservoir for viruses, and possibly facilitate their brain penetration.[62]

A number of nonautoimmune/infectious CNS disorders, including trauma and neurodegenerative diseases, present a pathophysiological inflammatory component. In these diseases, the CP–CSF system also seems to play a role.

Inflammation in Neurodegeneration in Acute and Neurodegenerative Disease

Chronic neuroinflammation is largely associated with neurodegenerative diseases like multiple sclerosis and AD. The term, "neuroinflammation," encompasses several molecular and cellular modifications without a clear, unified definition among the various brain diseases and injuries.[63] However, it is important to realize that, even if they share some of the same molecular players, neuroinflammation is distinct from peripheral inflammation particularly due to the involvement of microglia and astrocytes, which are cells specific to the CNS.[64,65] Since the inflammatory process may differ from organ to organ,[65] Graeber et al. recently drew the attention to the potential danger of using the two terminologies "neuroinflammation" and "inflammation" interchangeably.[64] As discussed earlier, the immune privilege of the brain was revised with the discovery that CSF is patrolled by activated T cells in physiological condition.[66] Interestingly, T cells after crossing the CP epithelium and trafficking in CSF were shown to support brain plasticity, in health as well as in response to CNS trauma.[2] This interesting result underlines the question whether inflammation is beneficial and deleterious on the final outcome in acute and neurodegenerative diseases.[2]

It is interesting to note that CP is a potential selective gate for the immune cell migration in physiological and after acute CNS trauma.[2] After spinal cord injury in rodents, a migration of leukocytes through the CP and ventricles, that is, far from the lesion site, has been observed.[67,68] The composition of the CSF is mostly immunosuppressive with the presence of cytokines such as IL-13 and TGF-β2.[68] Altogether, these observations suggest that following acute brain injury migration of leukocytes in CSF would have a role in ensuring the skewing capacity of these cells toward an anti-inflammatory phenotype. These cells would be involved in resolution of the local inflammation observed after CNS injury.[2]

The entry of the immune cells was proposed in postcapillary venules, where the molecular signaling has been found to facilitate the trafficking of the cells in pathological conditions such as autoimmune diseases exemplified by EAE.[18] In addition to this cerebrovascular route, the CP is also a route for the immune cells to penetrate in the CNS. As discussed earlier, the brain–CSF barrier does not exhibit any glia limitans barriers, and barrier properties are limited to tight junctions at the epithelium level in the choroid plexus.[69]

In addition to T cells, neutrophils and monocytes are also entering the CSF after acute parenchymal injury.[55,67,68,70] Trafficking of the immune cells is correlated with arrest of the local neuroinflammatory response, reduction of proinflammatory cytokine levels, and local elevation of neurotrophic factors.

To summarize, the local neuroinflammation with microglia and astrocyte activation could have a beneficial role by limiting the damages caused to the CNS after brain injury. Recruitment of immune cells at the CP, remotely from the lesion, and their further migration, could be part of the maturation of these cells toward an anti-inflammatory phenotype in order to stop the local neuroinflammation induced by brain injury. It is not yet very well understood how this concept could be applied for neurodegenerative diseases such as AD. However, relevant to AD, the importance of the immune system for the pathophysiology of the disease has been shown in murine models of AD, in which conditional ablation and reconstitution strategies by blood-borne macrophages attenuated the Aβ plaque formations.[71–73]

PART 3: CSF/ISF PROCESS IN NEURODEGENERATIVE DISEASES: PATHOPHYSIOLOGY AND BIOMARKERS

CSF and ISF Flow in Clearance of Metabolites: Involvement in Neurodegenerative Diseases

Dysfunctions of the choroid plexus and CSF formation have been described a long time ago.[74,75] The relevance of these dysfunctions to the pathophysiology of neurodegenerative diseases has been discussed since then. Several research groups proposed a potential role for CSF turnover and choroid plexus in late onset of AD.[30] It is interesting to think that the decrease of the CSF flow could be part of the onset of the neurodegenerative disease. For AD, the older demonstration that CSF plays a role in Aβ distribution and clearance from the brain,[37] and the recent publications from Nedergaard's group on the glymphatic system and its role in clearance of Aβ[45,76] are in support of this hypothesis formulated from the clinical observations.[30] Along the idea of the importance of the CSF–ISF flow for the clearance of Aβ, AQP4, which is present in all astrocyte end-feet in contact to all cerebral blood vessels,[20] has been proposed to have a role.[45] It has been shown that Aβ peptide clearance is significantly impaired in the perivascular space in the absence of AQP4 in the astrocyte end-feet.[45] Aβ accumulation around the blood vessels has been accompanied with changes in AQP4 expression in mouse models of AD,[77] and AD patients.[77,78] In a model of vascular amyloid, astrocytic AQP4 expression is significantly decreased[77] and a similar observation was made 2 months after juvenile TBI[63] associated with an increase in the Aβ load. All together these results suggest that a dysfunctional exchange between ISF and CSF due to a decrease of the AQP4 expression may contribute to the long-term accumulation of wastes around neurons. The presence of the wastes would induce local neuroinflammation with activation of microglia and astrocytes, and then neuronal cell death. This potential process in neurodegenerative diseases strongly suggests that glia dysfunctions caused by astrocyte phenotypic changes could be the starting point of the pathophysiology. In support of this hypothesis, a recent work reported that a decrease of the potassium rectifying channel 4.1 (Kir4.1) precedes dysfunctions of synaptic transmission and cell death in a rodent model of Huntington.[79] All these new data strongly suggest that CSF/ISF exchange dysfunctions occur before the neurologic symptoms appear. How such dysfunction is related to the primary alteration of CP morphology and secretory function observed in AD[80] remains to be established in detail. Because the CP also fulfills a number of other functions besides CSF secretion, including a role in brain detoxification, transport of biologically active molecules, and neuroimmune surveillance (see Part 1), it will be of interest to assess the status of these functions in degenerative diseases including AD, and their implication in the pathophysiological processes.

CSF: A Window on the Brain Diseases – Biomarkers for the Neurodegenerative Diseases

While only a limited number of studies address the role of CP and CSF flows in the pathophysiological process of neurodegenerative diseases, most of the works oriented toward CSF analysis have focused on the goal to find biomarkers in CSF for diagnosis of those neurodegenerative diseases. It is widely accepted that neurodegenerative diseases start with a preclinical stage beginning 10–20 years before the onset of the clinical symptoms. There is a strong need to develop CSF biomarkers, notwithstanding the collection of CSF is somewhat invasive with lumbar puncture. In parallel, neuroimaging with magnetic resonance imaging (MRI) and positron emission tomography (PET) are developing new modalities such as fMRI showing differences in connectivity between areas.[81] In PET-scan, a team in Pittsburgh developed the first radioligand to fibrillar Aβ, called Pittsburgh Compound B (PiB).[82] More recently, various amyloid imaging radioligands became available that utilize ^{18}F instead ^{11}C, facilitating its use in center with no cyclotron on site.[83] However, neuroimaging is still expensive and development of a biomarker found in the CSF and, better, in blood would be more suitable for several neurodegenerative diseases.

With regard to such development, several groups focused on the CSF as it reflects directly brain internal milieu. Biomarkers to be used as an indicator of the neurodegenerative diseases have to be measurable factor specific to the disease and reflecting the development of the pathophysiology. For AD, there are three core CSF biomarkers $Aβ_{42}$, total-tau (T-tau), and phospho-tau$_{181}$ (p-tau). $Aβ_{42}$ is a 42 amino acid peptide formed from the processing of amyloid precursor protein (APP). $Aβ_{42}$ is known to be the most abundant in the formation of amyloid plaques, and $Aβ_{42}$ concentration in CSF is significantly decreased in patients diagnosed with AD.[84] $Aβ_{42}$ decrease correlates inversely with *in vivo* amyloid plaques measurement in PET-scan[85,86] and in postmortem.[87] $Aβ_{42}$ is also increased in CSF in other brain diseases such as trauma, stroke, and other neurodegenerative diseases.[88] The second set of biomarkers proposed for AD is the detection of T-tau and p-tau in the CSF. Tau is a microtubule-associated protein present in neurons, which is released by the neurons without cell death, as well as present in ISF after neuronal cell death.[89] Interestingly, T-tau and p-tau have been found significantly increased in AD, associated with disease progression and in mild cognitive impairment (MCI).[90-92] However, high levels of T-tau and p-tau in CSF are observed with other brain diseases such as stroke, traumatic brain injury, and other neurodegenerative disease such as frontotemporal dementia (FTD) and dementia with Lewy body (DLB).[88] The combination of the changes in CSF for $A\beta_{42}$, T-tau and p-tau could be a better predictor in AD.[88] In the absence of a biomarker that will perform satisfactorily on its own, there is growing research to identify and develop additional CSF biomarkers. This new development of biomarkers explores other processes such as neuroinflammation. YKL-40 is a new candidate biomarker and it is a factor released by the activated microglia.[93] Studies have shown an increase of YKL-40 in CSF of AD patients.[94,95] There are also several other interesting candidates such as visinin-like protein-1, neurogranin, and F2-isoprostane.[88] The biomarker quest in the CSF as well as in blood is also focused on the exosomes and miRNA. Exosomes are endogenously produced vesicles that have a diameter of about 40–120 nm, with a lipid bilayer membrane that allows separation of the internal content of the exosome from the external environment. One of the physiological roles of exosomes is the transfer of RNA and miRNA.[96] Briefly, miRNA can affect posttranscriptional gene expression by either inhibiting translation or causing degradation of the mRNAs that code for specific proteins.[97] Numerous miRNAs have been identified so far in mammals, and thus far, several research groups have identified tissue-specific miRNA in the brain compared to other organs of the mammalian body.[96] miRNA expressions have been documented to change after models of TBI, ischemia, and neurodegenerative disease,[96] where different sets of miRNA showed changes at different time points after the injury. These observations suggest that miRNA also are interesting new targets for eventual biomarker factor in patients suffering from neurodegenerative diseases.

CONCLUSIONS

For several years the research in neurodegenerative diseases has been focusing on neuronal dysfunctions, overlooking the other cells and systems present in the brain. Due to the absence of success in the development of new drugs to cure or mitigate the progression of the disease, there is a new interest in considering the neuron as part of a larger system also comprising glia, brain vasculature, and ISF/CSF compartment. Recent studies point to dysfunctions of the CP and of CSF flow that could precede the neurological symptoms in neurodegenerative diseases. In addition, the CP presents a unique place of traffic for immune cells. It is interesting to underline that in neurodegenerative diseases, the migration of these immune cells in CSF seems to confer a beneficial effect by giving the capabilities to stop the local neuroinflammation. All together, these recent reports reinforce the idea that specific biomarkers can be found in CSF, which represents a window for the brain milieu.

Acknowledgments

The authors thank Dr N. Strazielle for fruitful discussion and bibliographical search. Supported by ANR-10-IBHU-0003 (JFGE), R01HD061946 (JB), Fondation des Gueules cassées (JB).

References

1. Pop V, Badaut J. A neurovascular perspective for long-term changes after brain trauma. *Transl Stroke Res.* 2011;2(4):533–545.
2. Schwartz M, Baruch K. The resolution of neuroinflammation in neurodegeneration: leukocyte recruitment via the choroid plexus. *EMBO J.* 2014;33(1): 7–22.
3. Davson H, Segal MB. *Physiology of the CSF and the Blood–Brain Barriers.* Boca Raton, New York, London, Tokyo: CRC Press; 1996.
4. Catala M. Embryonic and fetal development of structures associated with the cerebro-spinal fluid in man and other species. Part I: The ventricular system, meninges and choroid plexuses. *Arch Anat Cytol Pathol.* 1998;46(3):153–169.
5. Keep RF, Jones HC. A morphometric study on the development of the lateral ventricle choroid plexus, choroid plexus capillaries and ventricular ependyma in the rat. *Brain Res Dev Brain Res.* 1990;56(1):47–53.
6. Szmydynger-Chodobska J, Chodobski A, Johanson CE. Postnatal developmental changes in blood flow to choroid plexuses and cerebral cortex of the rat. *Am J Physiol.* 1994;266(5):1488–1492.
7. Keep RF, Ennis SR, Xiang J. *The Blood–CSF Barrier and Cerebral Ischemia.* Boca Raton, New York, London, Tokyo: CRC Press; 2005.
8. Falangola MF, Hanly A, Galvao-Castro B, Petito CK. HIV infection of human choroid plexus: a possible mechanism of viral entry into the CNS. *J Neuropathol Exp Neurol.* 1995;54(4):497–503.
9. McMenamin PG, Wealthall RJ, Deverall M, Cooper SJ, Griffin B. Macrophages and dendritic cells in the rat meninges and choroid plexus: three-dimensional localisation by environmental scanning electron microscopy and confocal microscopy. *Cell Tissue Res.* 2003;313(3):259–269.

10. Schmitt C, Strazielle N, Ghersi-Egea JF. Brain leukocyte infiltration initiated by peripheral inflammation or experimental autoimmune encephalomyelitis occurs through pathways connected to the CSF-filled compartments of the forebrain and midbrain. *J Neuroinflam.* 2012;9:187.

11. Serot JM, Bene MC, Foliguet B, Faure GC. Monocyte-derived IL-10-secreting dendritic cells in choroid plexus epithelium. *J Neuroimmunol.* 2000;105(2):115–119.

12. Serot JM, Foliguet B, Bene MC, Faure GC. Intraepithelial and stromal dendritic cells in human choroid plexus. *Hum Pathol.* 1998;29(10):1174–1175.

13. Nataf S, Strazielle N, Hatterer E, Mouchiroud G, Belin MF, Ghersi-Egea JF. Rat choroid plexuses contain myeloid progenitors capable of differentiation toward macrophage or dendritic cell phenotypes. *Glia.* 2006;54(3):160–171.

14. Ling EA, Kaur C, Lu J. Origin, nature, and some functional considerations of intraventricular macrophages, with special reference to the epiplexus cells. *Microsc Res Tech.* 1998;41(1):43–56.

15. Peters A, Palay LD, Webster H. *The Fine Structure of the Nervous System. Neurons and their Supporting Cells.* 3rd ed. New York Oxford: Oxford University Press; 1991.

16. Welch K, Sadler K. Permeability of the choroid plexus of the rabbit to several solutes. *Am J Physiol.* 1966;210(3):652–660.

17. Kratzer I, Vasiljevic A, Rey C, et al. Complexity and developmental changes in the expression pattern of claudins at the blood-CSF barrier. *Histochem Cell Biol.* 2012;138(6):861–879.

18. Engelhardt B, Coisne C. Fluids and barriers of the CNS establish immune privilege by confining immune surveillance to a two-walled castle moat surrounding the CNS castle. *Fluids Barriers CNS.* 2011;8(1):4.

19. Damkier HH, Brown PD, Praetorius J. Epithelial pathways in choroid plexus electrolyte transport. *Physiology (Bethesda).* 2010;25(4):239–249.

20. Badaut J, Fukuda AM, Jullienne A, Petry KG. Aquaporin and brain diseases. *Biochim Biophys Acta.* 2014;1840(5):1554–1565.

21. Brown PD, Davies SL, Speake T, Millar ID. Molecular mechanisms of cerebrospinal fluid production. *Neuroscience.* 2004;129(4):955–968.

22. Strazielle N, Khuth ST, Ghersi-Egea JF. Detoxification systems, passive and specific transport for drugs at the blood–CSF barrier in normal and pathological situations. *Adv Drug Deliv Rev.* 2004;56(12):1717–1740.

23. Strazielle N, Ghersi-Egea JF. Physiology of blood–brain interfaces in relation to brain disposition of small compounds and macromolecules. *Mol Pharma.* 2013;10(5):1473–1491.

24. Ghersi-Egea JF, Strazielle N, Murat A, Jouvet A, Buenerd A, Belin MF. Brain protection at the blood-cerebrospinal fluid interface involves a glutathione-dependent metabolic barrier mechanism. *J Cereb Blood Flow Metab.* 2006;26(9):1165–1175.

25. Strazielle N, Ghersi-Egea JF. Demonstration of a coupled metabolism-efflux process at the choroid plexus as a mechanism of brain protection toward xenobiotics. *J Neurosci.* 1999;19(15):6275–6289.

26. Kratzer I, Liddelow SA, Saunders NR, et al. Developmental changes in the transcriptome of the rat choroid plexus in relation to neuroprotection. *Fluids Barrier CNS.* 2013;10:25.

27. Redzic ZB, Segal MB. The structure of the choroid plexus and the physiology of the choroid plexus epithelium. *Adv Drug Deliv Rev.* 2004;56(12): 1695–1716.

28. Schmitt C, Strazielle N, Richaud P, Bouron A, Ghersi-Egea JF. Active transport at the blood-CSF barrier contributes to manganese influx into the brain. *J Neurochem.* 2011;117(4):747–756.

29. Chodobski A, Silverberg G, Szmydynger-Chodobska J. The role of the choroid plexus in transport and production of polypeptides. In: Zeng W, Chodobski A, eds. *The Blood–Cerebrospinal Fluid Barrier.* CRC Press, Boca Raton, FL, USA; 2005:237–274.

30. Serot JM, Zmudka J, Jouanny P. A possible role for CSF turnover and choroid plexus in the pathogenesis of late onset Alzheimer's disease. *J Alzheimers Dis.* 2012;30(1):17–26.

31. Monkkonen KS, Hakumaki JM, Hirst RA, et al. Intracerebroventricular antisense knockdown of G alpha i2 results in ciliary stasis and ventricular dilatation in the rat. *BMC Neurosci.* 2007;8:26.

32. Ghersi-Egea JF, Finnegan W, Chen JL, Fenstermacher JD. Rapid distribution of intraventricularly administered sucrose into cerebrospinal fluid cisterns via subarachnoid velae in rat. *Neuroscience.* 1996;75(4):1271–1288.

33. Iliff JJ, Lee H, Yu M, et al. Brain-wide pathway for waste clearance captured by contrast-enhanced MRI. *J Clin Invest.* 2013;123(3):1299–1309.

34. Bock NA, Paiva FF, Nascimento GC, Newman JD, Silva AC. Cerebrospinal fluid to brain transport of manganese in a non-human primate revealed by MRI. *Brain Res.* 2008;1198:160–170.

35. Johnson AK, Gross PM. Sensory circumventricular organs and brain homeostatic pathways. *FASEB J.* 1993;7(8):678–686.

36. Langlet F, Mullier A, Bouret SG, Prevot V, Dehouck B. Tanycyte-like cells form a blood–cerebrospinal fluid barrier in the circumventricular organs of the mouse brain. *J Comp Neurol.* 2013;521(15):3389–3405.

37. Ghersi-Egea JF, Gorevic PD, Ghiso J, Frangione B, Patlak CS, Fenstermacher JD. Fate of cerebrospinal fluid-borne amyloid beta-peptide: rapid clearance into blood and appreciable accumulation by cerebral arteries. *J Neurochem.* 1996;67(2):880–883.

38. Roberts J, Kahle MP, Bix GJ. Perlecan and the blood–brain barrier: beneficial proteolysis? *Front Pharmacol.* 2012;3:155.

39. Wolak DJ, Thorne RG. Diffusion of macromolecules in the brain: implications for drug delivery. *Mol Pharm.* 2013;10(5):1492–1504.

40. Cserr HF, Ostrach LH. Bulk flow of interstitial fluid after intracranial injection of blue dextran 2000. *Exp Neurol.* 1974;45(1):50–60.

41. Ichimura T, Fraser PA, Cserr HF. Distribution of extracellular tracers in perivascular spaces of the rat brain. *Brain Res.* 1991;545(1–2):103–113.

42. Abbott NJ. Evidence for bulk flow of brain interstitial fluid: significance for physiology and pathology. *Neurochem Int.* 2004;45(4):545–552.

43. Yamada S, DePasquale M, Patlak CS, Cserr HF. Albumin outflow into deep cervical lymph from different regions of rabbit brain. *Am J Physiol.* 1991;261 (4 Pt 2):1197–1204.

44. Rennels ML, Gregory TF, Blaumanis OR, Fujimoto K, Grady PA. Evidence for a "paravascular" fluid circulation in the mammalian central nervous system,

provided by the rapid distribution of tracer protein throughout the brain from the subarachnoid space. *Brain Res.* 1985;326(1):47–63.

45. Iliff JJ, Wang M, Liao Y, et al. A paravascular pathway facilitates CSF flow through the brain parenchyma and the clearance of interstitial solutes, including amyloid beta. *Sci Transl Med.* 2012;4(147):147ra111.

46. Xie L, Kang H, Xu Q, et al. Sleep drives metabolite clearance from the adult brain. *Science.* 2013;342(6156):373–377.

47. Yang L, Kress BT, Weber HJ, et al. Evaluating glymphatic pathway function utilizing clinically relevant intrathecal infusion of CSF tracer. *J Transl Med.* 2013;11:107.

48. Strittmatter WJ. Bathing the brain. *J Clin Invest.* 2013;123(3):1013–1015.

49. Iliff JJ, Wang M, Zeppenfeld DM, et al. Cerebral arterial pulsation drives paravascular CSF-interstitial fluid exchange in the murine brain. *J Neurosci.* 2013;33(46):18190–18199.

50. Binder DK, Papadopoulos MC, Haggie PM, Verkman AS. *In vivo* measurement of brain extracellular space diffusion by cortical surface photobleaching. *J Neurosci.* 2004;24(37):8049–8056.

51. Wen H, Nagelhus EA, Amiry-Moghaddam M, Agre P, Ottersen OP, Nielsen S. Ontogeny of water transport in rat brain: postnatal expression of the aquaporin-4 water channel. *Eur J Neurosci.* 1999;11(3):935–945.

52. Ransohoff RM, Engelhardt B. The anatomical and cellular basis of immune surveillance in the central nervous system. *Nat Rev Immunol.* 2012;12(9):623–635.

53. Giunti D, Borsellino G, Benelli R, et al. Phenotypic and functional analysis of T cells homing into the CSF of subjects with inflammatory diseases of the CNS. *J Leukoc Biol.* 2003;73(5):584–590.

54. Kivisakk P, Mahad DJ, Callahan MK, et al. Human cerebrospinal fluid central memory CD4+ T cells: evidence for trafficking through choroid plexus and meninges via P-selectin. *Proc Natl Acad Sci USA.* 2003;100(14):8389–8394.

55. Carrithers MD, Visintin I, Kang SJ, Janeway Jr CA. Differential adhesion molecule requirements for immune surveillance and inflammatory recruitment. *Brain.* 2000;123(Pt 6):1092–1101.

56. Piccio L, Rossi B, Scarpini E, et al. Molecular mechanisms involved in lymphocyte recruitment in inflamed brain microvessels: critical roles for P-selectin glycoprotein ligand-1 and heterotrimeric G(i)-linked receptors. *J Immunol.* 2002;168(4):1940–1949.

57. Petito CK, Adkins B. Choroid plexus selectively accumulates T-lymphocytes in normal controls and after peripheral immune activation. *J Neuroimmunol.* 2005;162(1–2):19–27.

58. Reboldi A, Coisne C, Baumjohann D, et al. C-C chemokine receptor 6-regulated entry of TH-17 cells into the CNS through the choroid plexus is required for the initiation of EAE. *Nat Immunol.* 2009;10(5):514–523.

59. Brown DA, Sawchenko PE. Time course and distribution of inflammatory and neurodegenerative events suggest structural bases for the pathogenesis of experimental autoimmune encephalomyelitis. *J Comp Neurol.* 2007;502(2):236–260.

60. Kutzelnigg A, Lassmann H. Cortical lesions and brain atrophy in MS. *J Neurol Sci.* 2005;233(1–2):55–59.

61. Ubogu EE, Cossoy MB, Ransohoff RM. The expression and function of chemokines involved in CNS inflammation. *Trends Pharmacol Sci*. 2006;27(1):48–55.

62. Hanly A, Petito CK. HLA-DR-positive dendritic cells of the normal human choroid plexus: a potential reservoir of HIV in the central nervous system. *Hum Pathol*. 1998;29(1):88–93.

63. Fukuda AM, Pop V, Spagnoli D, Ashwal S, Obenaus A, Badaut J. Delayed increase of astrocytic aquaporin 4 after juvenile traumatic brain injury: possible role in edema resolution? *Neuroscience*. 2012;222:366–378.

64. Graeber MB, Li W, Rodriguez ML. Role of microglia in CNS inflammation. *FEBS Lett*. 2011;585(23):3798–3805.

65. Kelley KW, Dantzer R. Alcoholism and inflammation: neuroimmunology of behavioral and mood disorders. *Brain Behav Immun*. 2011;25(suppl 1):S13–20.

66. Engelhardt B, Ransohoff RM. The ins and outs of T-lymphocyte trafficking to the CNS: anatomical sites and molecular mechanisms. *Trends Immunol*. 2005;26(9):485–495.

67. Kunis G, Baruch K, Rosenzweig N, et al. IFN-gamma-dependent activation of the brain's choroid plexus for CNS immune surveillance and repair. *Brain*. 2013;136(Pt 11):3427–3440.

68. Shechter R, Miller O, Yovel G, et al. Recruitment of beneficial M2 macrophages to injured spinal cord is orchestrated by remote brain choroid plexus. *Immunity*. 2013;38(3):555–569.

69. Redzic Z. Molecular biology of the blood–brain and the blood–cerebrospinal fluid barriers: similarities and differences. *Fluids Barriers CNS*. 2011;8(1):3.

70. Szmydynger-Chodobska J, Strazielle N, Gandy JR, et al. Posttraumatic invasion of monocytes across the blood–cerebrospinal fluid barrier. *J Cereb Blood Flow Metab*. 2012;32(1):93–104.

71. Simard AR, Soulet D, Gowing G, Julien JP, Rivest S. Bone marrow-derived microglia play a critical role in restricting senile plaque formation in Alzheimer's disease. *Neuron*. 2006;49(4):489–502.

72. Butovsky O, Kunis G, Koronyo-Hamaoui M, Schwartz M. Selective ablation of bone marrow-derived dendritic cells increases amyloid plaques in a mouse Alzheimer's disease model. *Eur J Neurosci*. 2007;26(2):413–416.

73. Town T, Laouar Y, Pittenger C, et al. Blocking TGF-beta-Smad2/3 innate immune signaling mitigates Alzheimer-like pathology. *Nat Med*. 2008;14(6):681–687.

74. Sohn RS, Siegel BA, Gado M, Torack RM. Alzheimer's disease with abnormal cerebrospinal fluid flow. *Neurology*. 1973;23(10):1058–1065.

75. Coblentz JM, Mattis S, Zingesser LH, Kasoff SS, Wisniewski HM, Katzman R. Presenile dementia. Clinical aspects and evaluation of cerebrospinal fluid dynamics. *Arch Neurol*. 1973;29(5):299–308.

76. Kress BT, Iliff JJ, Xia M, et al. Impairment of paravascular clearance pathways in the aging brain. *Ann Neurol*. 2014;76(6):845–861.

77. Wilcock DM, Vitek MP, Colton CA. Vascular amyloid alters astrocytic water and potassium channels in mouse models and humans with Alzheimer's disease. *Neuroscience*. 2009;159(3):1055–1069.

78. Moftakhar P, Lynch MD, Pomakian JL, Vinters HV. Aquaporin expression in the brains of patients with or without cerebral amyloid angiopathy. *J Neuropathol Exp Neurol*. 2010;69(12):1201–1209.

79. Tong X, Ao Y, Faas GC, et al. Astrocyte Kir4.1 ion channel deficits contribute to neuronal dysfunction in Huntington's disease model mice. *Nat Neurosci*. 2014;17(5):694–703.

80. Serot JM, Bene MC, Foliguet B, Faure GC. Morphological alterations of the choroid plexus in late-onset Alzheimer's disease. *Acta Neuropathol*. 2000;99(2):105–108.

81. Sheline YI, Raichle ME. Resting state functional connectivity in preclinical Alzheimer's disease. *Biol Psychiatr*. 2013;74(5):340–347.

82. Wang Y, Klunk WE, Debnath ML, et al. Development of a PET/SPECT agent for amyloid imaging in Alzheimer's disease. *J Mol Neurosci*. 2004;24(1): 55–62.

83. Mathis CA, Mason NS, Lopresti BJ, Klunk WE. Development of positron emission tomography beta-amyloid plaque imaging agents. *Semin Nucl Med*. 2012;42(6):423–432.

84. Blennow K, Hampel H, Weiner M, Zetterberg H. Cerebrospinal fluid and plasma biomarkers in Alzheimer disease. *Nat Rev*. 2010;6(3):131–144.

85. Fagan AM, Mintun MA, Mach RH, et al. Inverse relation between *in vivo* amyloid imaging load and cerebrospinal fluid Abeta42 in humans. *Ann Neurol*. 2006;59(3):512–519.

86. Forsberg A, Almkvist O, Engler H, Wall A, Langstrom B, Nordberg A. High PIB retention in Alzheimer's disease is an early event with complex relationship with CSF biomarkers and functional parameters. *Curr Alzheimer Res*. 2010;7(1):56–66.

87. Strozyk D, Blennow K, White LR, Launer LJ. CSF Abeta 42 levels correlate with amyloid-neuropathology in a population-based autopsy study. *Neurology*. 2003;60(4):652–656.

88. Rosen C, Hansson O, Blennow K, Zetterberg H. Fluid biomarkers in Alzheimer's disease – current concepts. *Mol Neurodegener*. 2013;8:20.

89. Yamada K, Cirrito JR, Stewart FR, et al. *In vivo* microdialysis reveals age-dependent decrease of brain interstitial fluid tau levels in P301S human tau transgenic mice. *J Neurosci*. 2011;31(37):13110–13117.

90. Buerger K, Ewers M, Pirttila T, et al. CSF phosphorylated tau protein correlates with neocortical neurofibrillary pathology in Alzheimer's disease. *Brain*. 2006;129(Pt 11):3035–3041.

91. Buerger K, Otto M, Teipel SJ, et al. Dissociation between CSF total tau and tau protein phosphorylated at threonine 231 in Creutzfeldt-Jakob disease. *Neurobiol Aging*. 2006;27(1):10–15.

92. Blom ES, Giedraitis V, Zetterberg H, et al. Rapid progression from mild cognitive impairment to Alzheimer's disease in subjects with elevated levels of tau in cerebrospinal fluid and the APOE epsilon4/epsilon4 genotype. *Dement Geriatr Cogn Disord*. 2009;27(5):458–464.

93. Bonneh-Barkay D, Bissel SJ, Wang G, et al. YKL-40, a marker of simian immunodeficiency virus encephalitis, modulates the biological activity of basic fibroblast growth factor. *Am J Pathol*. 2008;173(1):130–143.

94. De Meyer G, Shapiro F, Vanderstichele H, et al. Diagnosis-independent Alzheimer disease biomarker signature in cognitively normal elderly people. *Arch Neurol*. 2010;67(8):949–956.

95. Olsson B, Hertze J, Ohlsson M, et al. Cerebrospinal fluid levels of heart fatty acid binding protein are elevated prodromally in Alzheimer's disease and vascular dementia. *J Alzheimers Dis*. 2013;34(3):673–679.
96. Fukuda AM, Badaut J. siRNA treatment: "A Sword-in-the-Stone" for acute brain injuries. *Genes*. 2013;4(3):435–456.
97. Carthew RW, Sontheimer EJ. Origins and mechanisms of miRNAs and siRNAs. *Cell*. 2009;136(4):642–655.

CHAPTER

9

Delivery Considerations for Targeting the Choroid Plexus–Cerebrospinal Fluid Route

Thomas C. Chen

Department of Neurological Surgery, Keck School of Medicine, University of Southern California, Los Angeles, CA, USA

The Choroid Plexus and Cerebrospinal Fluid. http://dx.doi.org/10.1016/B978-0-12-801740-1.00009-3

INTRODUCTION

The choroid plexus–cerebrospinal fluid (CSF) barrier is formed by the choroid plexus epithelium. The choroid plexus is a secretory epithelial tissue made up of the choroid plexus epithelial cell, stroma, and fenestrated endothelial cells. The choroid plexus has a multitude of important functions including manufacturing of CSF, homeostatic functions related to stabilizing the interstitial environment of neurons, and kidney, liver, and immune type functions. In the fetus, the choroid plexus is involved in the development of the periventricular neurogenic zones. In later stages of life, the choroid plexus is challenged to potential toxic molecules where the turnover of CSF and clearance of potentially toxic materials are keys to maintaining neuronal lifespan. In health, the choroid plexus is key to delivering micronutrients, growth factors, and neurotrophins to neuronal networks via the CSF from systemic blood delivered to the leaky microvasculature of the choroid plexus.

Because the choroid plexus is so essential to brain viability, drug delivery to the choroid plexus– CSF barrier should be examined much more closely. To date, drug delivery to the brain has focused on blood–brain barrier (BBB). Drug delivery to and via the choroid plexus may occur in four different fashions: (1) From blood to choroid plexus epithelium to CSF, resulting in delivery of small molecules, (i.e. micronutrients, neurotrophins, neuropeptides, growth factors to many periventricular targets such as caudate, hippocampus, circumventricular organs, hypothalamus, pia-glia, and arachnoid membranes); (2) targeting the choroid plexus epithelium itself, influencing secretion of CSF and other organic ions secreted by the choroid plexus; (3) from CSF to choroid plexus epithelium to the brain interstitial fluid; and (4) from CSF to choroid plexus to the systemic circulation. Although all these delivery routes involve the choroid plexus, the ease of delivery is dramatically different. For instance, if the desired access is the choroid plexus epithelium, then its access is easy, and may be obtained via an intraventricular access. However, if the desired target is intraparenchymal, then the ability to get to the interstitial fluid bathing the brain parenchyma is much more difficult.

In this chapter, we will first define the anatomy of the choroid plexus and how it separates CSF from the normal brain. We will then discuss, the advantages and disadvantages of delivery to the brain via the choroid plexus–CSF barrier. Third, we will discuss the future technology that may be used to deliver to the choroid plexus. And finally, we will discuss the potential disease states to treat and the concept of delivery, that is, what to target, what to deliver, and how to deliver, will be highlighted.

ANATOMICAL CONSIDERATIONS

The choroid plexus is made up of three layers: the choroid plexus epithelial cell (CPE), stroma, and endothelial cell. It is suspended at multiple loci in the lateral ventricle, III ventricle, and IV ventricle. The choroid plexus is very well perfused, receiving blood from internal carotid arteries and the vertebral artery. In the lateral ventricles, the choroid plexuses are supplied by the anterior circulation (anterior choroidal artery-branch of middle cerebral artery) and posterior circulation (posterior choroidal artery-branch of vertebral artery). The choroid plexus of the third ventricle is supplied by the posterior choroidal artery. The choroid plexus of the fourth ventricle is supplied by the posterior circulation (branches of AICA – anterior inferior cerebellar artery; and PICA –posterior inferior cerebellar artery). The choroid plexuses are perfused with blood at a rate of 4 mL/min/g of choroid plexus tissue, which is approximately 10 times higher than the rate of blood supply to the brain parenchyma.[1] Innervation of the choroid plexus is via the sympathetic and parasympathetic systems. The sympathetic system controls blood flow to the choroid plexus via the fibers arising from the superior cervical ganglion. The parasympathetic system reduces CSF production.[2]

The CPE is on the apical side of the choroid plexus, facing the ventricle; the stroma and endothelial cell on the basolateral surface. The CPE folds itself into a cauliflower like structure, protruding into the lateral, III, and IV ventricles. CPE cells also have multiple cilia protruding, expanding the actual surface area available for absorption. The CPE cells are responsible for the synthesis of CSF, which bathes and protects the brain. There is approximately 140 mL of CSF in the human brain, which fills four ventricles namely lateral ventricles, III ventricle, and ventricle (20 mL); spinal subarachnoid space (30 mL); and cranial subarachnoid space (90 mL). Recently, it has been shown that the CSF communicates with the brain parenchyma via the subarachnoid "paravascular" spaces, mixes with CSF, and is cleared via the perivenous drainage. The stroma is past the CPE. Like all other connective tissues, it is being increasingly recognized for its importance in establishment of the microenvironment. The choroid plexus stroma has been increasingly recognized for its role within the immune microenvironment. Lastly, the endothelial cell within the choroid plexus is fenestrated. It serves as a route for high flow output from the blood stream into the CSF. This high flow output enables small molecules to easily get from the bloodstream into the CSF.

PROS AND CONS OF DRUG DELIVERY VIA THE CHOROID PLEXUS–CSF BARRIER

As stated earlier, there are four potential avenues of delivery to the brain involving the choroid plexus–CSF barrier. In this section, we will discuss each delivery route, emphasizing on the advantages and disadvantages of each approach.

In the first approach, drug delivery to the CSF, comes with a goal of targeting periventricular brain, ependymal, circumventricular organs, or leptomeningeal cancer cells. Direct delivery either intravenously or intraventricularly is the route of choice. In the intravenous approach, drug is delivered from the systemic circulation to the microvasculature of the choroid plexus. Therefore, direct delivery to the choroid plexus microvasculature is very rapid and unimpeded. Because the choroid plexus vasculature is fenestrated, drug transfer from blood to CSF occurs much more readily across the choroid plexus–CSF barrier compared to the BBB. Virtually all small and large molecules in blood penetrate into the CSF, at a rate inversely proportional to the molecular weight of the substance. It should be emphasized that drug delivery into the CSF does not equate with BBB permeability. Most drugs will get into the CSF, as the fenestrated endothelium and choroid plexus epithelium are much different from the BBB. This difference in delivery is highlighted by the concentration of the anti-HIV protease azidothymidine (AZT) after intravenous delivery. AZT can be easily detected in the CSF; however, penetration in the brain parenchyma is extremely limited secondary to poor BBB and the efflux mechanisms pumping it out at the BBB.[3]

If the drug delivery is leptomeningeal (where cancer cells are located), the preferential route would be direct intraventricular via an Ommaya reservoir or via lumbar intrathecal delivery. In this route, the drug is delivered to the CSF in high levels, where the vast majority of it is concentrated. Since the penetration of the drug decreases exponentially with the distance from the CSF, a large proportion of the drug will be at the ependymal surface, where the cancer cells are located.[4] Chen et al. have delivered mesenchymal stromal cells to the CSF with the goal of delivering genes into the CSF for neurodegenerative diseases. The mesenchymal stromal cells would be used for supportive purposes in the treatment of neurodegenerative diseases such as amylotrophic lateral sclerosis (ALS).[5]

In the second approach, direct targeting of the CPE with the goal of changing its behavior, affecting transport of specific molecules, and affecting degree of CSF secretion also has certain advantages. Gonzalez et al.[6] have looked at the possibility of targeting the CPE by tropic ligands generated from a peptide library displayed on M13 bacteriophage and screening for ligands capable of internalizing into CPE by incubating phage with CP explants and recovering particles for targeting capacity. In this way, they generated three peptides that were capable of targeting CPE.[6] Baird et al. have examined the use of bacteriophage targeting of epidermal growth factor receptor (EGFR) expressed on choroid plexus cells with the goal of introducing genes into the choroid plexus.[7]

In the third approach, if the target is in the parenchyma of the brain, then targeting the choroid plexus of the brain for delivery is currently not going to be the most efficient route. Pardridge has continuously emphasized that drug concentration levels in the CSF do not equate to direct delivery to the brain parenchyma in human patients, and is not a measure of BBB permeability. The reason is simple. In the human brain, drugs penetrate the fenestrated endothelium of the choroid plexus easily, and are carried into the CSF, at a rate inversely related to the molecular weight of the substance. Once they are in the CSF, they can be carried via the CSF; however, even small molecules will be hampered from direct communication with the brain parenchyma via the glia limitans (a layer of interdigitating astrocytic processes with an overlying basement membrane). Therefore, direct delivery via convection-enhanced delivery using a catheter placed at the parenchymal target or bypassing the blood–brain barrier is going to be much more effective.

The fourth approach is delivery to the CSF, penetration of the choroid epithelium to the systemic circulation. Pardridge has equated this route of penetration as equivalent to a slow intravenous infusion. Following ICV injection of the drug, it is able to move along the CSF flow tracks (i.e., through ventricular system to foramens of Magendie and Luschka) to the arachnoid villi, and then to the general circulation, where it will have to enter the parenchyma via the BBB. This is illustrated by barbiturate injection into the CSF in dogs. The barbiturate exits the CSF, and reenters the brain parenchyma via the BBB similar to an intravenous dose of barbiturates.[8]

NOVEL METHODS FOR CSF DELIVERY AND CHOROID PLEXUS TARGETING

Targeting Choroid Plexus Epithelial Cells and Gene Therapy

Targeting choroid plexus epithelial cells to alter their behavior have been investigated by a number of researchers. Strategies have ranged from using bacteriophage synthesized ligand-mediated targeting of choroid plexus cells, bacteriophage expressing epidermal growth factor (EGF) targeting for gene delivery, and demonstration of esophageal-cancer-related gene-4 (ECGR-4) as a major site of gene expression in the choroid plexus. The use of bacteriophage to synthesize ligands to target the choroid plexus is a novel and interesting method of targeting the choroid plexus. This technique was developed by Gonzalez et al.,[6] who incubated the choroid plexus epithelium with the bacteriophage for 2 h at 37°C, and recovered particles with targeting capacity. From this library, three peptides were identified that had high binding affinity to choroid plexus epithelial cells.[9] Subsequently, bacteriophage engineered to express epidermal growth factor (EGF) was also shown to specifically target CPE, demonstrating that EGFR is overexpressed in CPE.[9] Demonstration of the tumor suppressor gene ECGR-4 in normal choroid epithelial cells, but not in choroid plexus carcinomas or malignant gliomas allows for targeting specificity for normal cells, and not cancer cells. These targeting techniques are all aimed at targeting the apical surface of the choroid plexus, with delivery via the CSF.

Metronomic Biofeedback Pump

The metronomic biofeedback pump (MBP; Pharmaco-kinesis) is a sophisticated implanted pump that was built for metronomic delivery of drugs with the capacity to monitor drug concentrations using absorption spectroscopy. The first indication for the MBP is in the treatment of leptomeningeal carcinomatosis. The pump has three components: (1) a ventricular catheter reservoir (inserted into the lateral ventricle), (2) a catheter (double lumen for drug delivery and CSF retrieval for analysis), and (3) the pump itself, which contains two 5 mL refillable reservoirs. Drug (currently methotrexate) is delivered into the CSF using one lumen of the catheter. Sampling of CSF will be performed in the other. Methotrexate concentration in the CSF is determined by a built-in spectrophotometer, which determines concentration by absorbance. The pump will be implanted in the lateral chest wall, with the catheter implanted in the lateral ventricle.[10]

Use of Immune Cells for Delivery

The use of immune cells for drug delivery is a novel potential vehicle to modulate the choroid plexus. Immune cells (macrophages, T cells, dendritic cells) normally have bidirectional transport between the systemic circulation and the CSF. The use of immune cells to deliver cytokines to the CSF via systemic delivery can be easily accomplished. Conversely, immune cells (i.e., dendritic cells) can potentially be delivered intraventricularly to elicit a more lasting antigen presenting response. Neuroinflammatory conditions, such as head trauma, multiple sclerosis, and infections, may all potentially be modulated by specific immune modulators that may be able to decrease the inflammatory response.[11,12]

Somatic Therapy with Engineered Choroid Plexus Cells

Somatic therapy with engineered choroid plexus cells is another technique that has been used to influence the choroid plexus. In certain disease states, such as Alzheimer's, the ability of choroid plexus cells to perform their duties of secretion of CSF and maintenance of CSF homeostasis is decreased. One problem with engineered choroid plexus cells is the difficulty in maintaining them in culture. Recently, Watanabe et al. have examined the role of bone morphogenetic protein 4 (BMP4) in the induction of choroid plexus epithelial fate from embryonic stem-cell-derived neuroepithelial progenitors.[13]

CLINICAL IMPLICATIONS OF DRUG DELIVERY TO CHOROID PLEXUS

This section will highlight four disease states that may be potentially treated by delivery to the choroid plexus.

Pseudotumor Cerebri

Pseudotumor cerebri is for the most part an idiopathic disease state seen in young or middle-aged obese women. It has also been associated with over-ingestion of vitamin A or use of oral contraceptives. It is associated with increased intracranial pressure (often with normal-sized ventricles) resulting in severe headaches or papilledema, and optic nerve swelling. Conservative treatment consists of the use of acetazolomide (Diamox) or weight loss in obese patients. Acetazolamide inhibits carbonic anhydrase in the choroid epithelial cells. It has been demonstrated to decrease CSF production in man.[14] As a drug that is given systemically, acetazolamide is able to be taken orally, be transported to the choroid epithelial cell, and inhibit carbonic anhydrase in the choroid epithelial cells. Carbonic anhydrase is one of the many ion pumps present in the choroid epithelial cell; inhibition of its activity inhibits CSF production.[15,16] Although acetazolamide is useful in decreasing CSF production, patients often develop resistance to its actions. As a result, a new generation of agents, such as receptors for arginine vasopressin (AVP) and atrial natriuretic peptide (ANP), in choroid plexus epithelial cells are targets for agents to reduce CSF formation.[17] Patients that are resistant to conservative treatment are treated with a CSF diversionary procedure depending on their symptoms. Patients with predominantly decreased vision are treated with optic nerve fenestration. Patients with predominantly headaches are treated with either a ventriculoperitoneal or lumboperitoneal shunt.

Alzheimer's Disease

Alzheimer's disease (AD) is a disease that will be increasing in numbers worldwide. It is estimated that there are currently approximately 30 million cases worldwide that will triple in the next 40 years. The etiology of AD has been associated with amyloid-B

(AB) deposition in the form of amyloid plaques, tau aggregation in the form of neurofibrillary tangles, and neuroinflammation resulting in neuronal and synaptic loss in certain brain regions such as the hippocampus. CSF levels of AB actually decrease (presumably by increased AB deposition in the brain), followed by increase in tau levels, and the neuronal protein VILIP-1. CSF levels of growth factors, neuroprotective peptides, and neuroinflammatory cytokines have all been actively investigated. Because of the profoundly altered composition of brain interstitial fluid composition there is a compromise in neurotransmission as well as neuronal metabolism. The choroid plexus epithelium and its role in secretion of CSF are impaired. Drugs that can be delivered to improve the function of the CPE, or the creation of new functional CPE via genetic engineering, will be useful in the treatment of Alzheimer's.[13]

Leptomeningeal Carcinomatosis

Leptomeningeal carcinomatosis is usually a late manifestation of systemic cancer. It involves seeding of carcinoma cells to the leptomeninges (pia and arachnoid). Patients typically present with headaches, cranial neuropathies, or dizziness. It is often fatal within 6–8 months after diagnosis, depending on the type of cancer. Current therapy is usually intraventricular, with delivery of either methotrexate of Depocyt (liposomal Ara-C) via an Ommaya reservoir. Delivery for this disease is via the first-approach delivery to the CSF, with eventual collection in the subarachnoid space.[10]

Choroid Plexus Papillomas

Choroid plexus papillomas are tumors of the choroid plexus and are usually pediatric tumors. They can be benign choroid plexus papillomas or malignant choroid plexus carcinomas. These tumors are best treated with complete surgical resection for benign choroid plexus papillomas. Choroid plexus carcinomas, however, even with radiographically complete resections need to be treated with adjunctive therapy such as radiation and/or chemotherapy. Choroid plexus tumors can be treated using similar targeting techniques that have been developed for the CPE. Gene targeting and bacteriophage ligand targeting are all techniques that should be equally applicable. The route of application is the second one, directly to the CPE.[9]

CONCLUSIONS

This chapter has emphasized the role of drug delivery to the choroid plexus–CSF barrier. There are four main routes of application: (1) delivery to the CSF from systemic administration or via external Ommaya reservoir or implanted intrathecal pump; (2) delivery to the CPE itself, using targeting techniques; (3) delivery to the parenchymal interstitial fluid via paravenous drainage; and (4) delivery to the systemic circulation via the choroid plexus. Traditional methods and new methods of drug targeting of these four routes of administration have been discussed. Finally, disease categories using these routes of administration have been discussed.

References

1. Keep RF, Jones HC. A morphometric study on the development of the lateral ventricle choroid plexus, choroid plexus capillaries and ventricular ependymal in the rat. *Brain Res Dev Brain Res*. 1990;56(1):47–53.
2. Damkier HH, Brown PD, Praetorius J. Cerebrospinal fluid secretion by the choroid plexus. *Physiol Rev*. 2013;93(4):1847–1892.
3. Dykstra KH, Arya A, Arriola DM, Bungay PM, Morrison PF, Dedrick RL. Microdialysis study of zidovudine (AZT) transport in rat brain. *J Pharmacol Exp Ther*. 1993;267(3):1227–1236.
4. Blasberg RG, Patlak C, Fenstermacher JD. Intrathecal chemotherapy: brain tissue profiles after ventriculocisternal perfusion. *J Pharmacol Exp Ther*. 1975;195(1):73–83.
5. Chen BK, Staff NP, Knight AM, et al. A safety study on intrathecal delivery of autologous mesenchymal stromal cells in rabbits directly supporting Phase I human trials. *Transfusion*. 2015;55(5):1013–1020.
6. Gonzalez AM, Leadbeater WE, Burg M, et al. Targeting choroid plexus epithelia and ventricular ependyma for drug delivery to the central nervous system. *BMC Neurosci*. 2011;12:4.
7. Baird A, Eliceiri BP, Gonzalez AM, Johanson CE, Leadbeater W, Stopa EG. Targeting the choroid plexus-CSF-brain nexus using peptides identified by phage display. *Methods Mol Biol*. 2011;686:483–498.
8. Aird RB. A study of intrathecal, cerebrospinal fluid-to-brain exchange. *Exp Neurol*. 1984;86(2):342–358.
9. Gonzalez AM, Podvin S, Leadbeater W. Epidermal Growth Factor Targeting of Bacteriophage to the Choroid Plexus for Gene Delivery to the Central Nervous System via Cerebrospinal Fluid. *Brain Res*. 2010;1359:1–13.
10. Chen TC, Napolitano GR, Adell F, Schönthal AH, Shachar Y. Development of the Metronomic Biofeedback Pump for leptomeningeal carcinomatosis: technical note. *J Neurosurg*. 2015;123(2):362–372.
11. Baruch K, Schwartz M. CNS-specific T cells shape brain function via the choroid plexus. *Brain Behav Immun*. 2013;34:11–16.

12. Johanson C, Stopa E, McMillan P, Roth D, Funk J, Krinke G. The distributional nexus of choroid plexus to cerebrospinal fluid, ependyma and brain: toxicologic/pathologic phenomena, periventricular destabilization, and lesion spread. *Toxicol Pathol*. 2011;39(1):186–212.

13. Watanabe M, Kang YJ, Davies LM, et al. BMP4 sufficiency to induce choroid plexus epithelial fate from embryonic stem cell-derived neuroepithelial progenitors. *J Neurosci*. 2012;32(45):15934–15945.

14. Rubin RC, Henderson ES, Ommaya AK, Walker MD, Rall DP. The production of cerebrospinal fluid in man and its modification by acetazolamide. *J Neurosurg*. 1966;25(4):430–436.

15. Vogh BP. The relation of choroid plexus carbonic anhydrase activity to cerebrospinal fluid formation: study of three inhibitors in cat with extrapolation to man. *J Pharmacol Exp Ther*. 1980;213(2):321–331.

16. Damkier HH, Brown PD, Praetorius J. Cerebrospinal fluid secretion by the choroid plexus. *Physiol Rev*. 2013;93(4):1847–1892.

17. Johanson CE, Preston JE, Chodobski A, et al. AVP V1 receptor-mediated decrease in Cl- efflux and increase in dark cell number in choroid plexus epithelium. *Am J Physiol*. 1999;276:C82–90.

Glossary

5-Bromo-2-deoxyuridine (BrdU) A synthetic nucleoside that is an analog of thymidine. BrdU is commonly used in the detection of proliferating cells in living tissues.

Anterior inferior cerebellar artery (AICA) An artery in the brain that supplies part of the cerebellum. It arises from the basilar artery at the level of the junction between the medulla oblongata and the pons in the brainstem.

Alpha subunit (Atp1a1) An integral membrane protein responsible for establishing and maintaining the electrochemical gradients of Na and K ions across the plasma membrane.

Alzheimer's disease A progressive degenerative disease of the brain that causes impairment of memory and dementia manifested by confusion, visual–spatial disorientation, impairment of language function progressing from anomia to fluent aphasia, inability to calculate, and deterioration of judgment; delusions and hallucinations may occur.

Amyloid Beta (Aβ) Peptides of 36–43 amino acids that are crucially involved in Alzheimer's disease as the main component of the amyloid plaques found in the brains of Alzheimer patients.

Amyloid plaques Amyloid is a general term for protein fragments that the body produces normally. Beta amyloid is a protein fragment snipped from an amyloid precursor protein (APP). In a healthy brain, these protein fragments are broken down and eliminated. In Alzheimer's disease, the fragments accumulate to form hard, insoluble plaques.

Anterior choroidal arteries (AChA) Artery in the brain that originates from the internal carotid artery and serves structures in the telencephalon, diencephalon, and mesencephalon.

Antigen presenting cell (APC) Also called an accessory cell, it is a cell that displays foreign antigens complexed with major histocompatibility complexes (MHCs) on their surfaces; these cells process antigens and present them to T cells.

Aquaporin-4 (AQP4) An aquaporin expressed in the basolateral cell membrane of principal collecting duct cells in the kidney and provide a pathway for water to exit these cells. AQP4 is also expressed in astrocytes and is upregulated by direct insult to the central nervous system (CNS).

Aquaporin A member of a family of transmembrane channel proteins found in epithelial membranes that serve to regulate transepithelial water movement in tissues involved in body fluid homeostasis.

Aqueduct cerebri See **cerebral aqueduct**

Aqueduct of Sylvius Within the mesencephalon (or midbrain), contains cerebrospinal fluid (CSF), and connects the third ventricle in the diencephalon to the fourth ventricle within the region of the mesencephalon and metencephalon, located dorsal to the pons and ventral to the cerebellum.

The Choroid Plexus and Cerebrospinal Fluid. http://dx.doi.org/10.1016/B978-0-12-801740-1.00019-6

Astrocyte One of the large neuroglia cells of nervous tissue, many of which form the blood–brain barrier and allow nutrients to be brought to the brain by dilation of blood vessels.

Astrocytic "end-feet" The part of the astrocyte that ensheathes blood vessels in the brain and are believed to provide structural integrity to the cerebral vasculature.

ATP binding cassette (ABC) Transmembrane proteins that utilize the energy of adenosine triphosphate (ATP) binding and hydrolysis to carry out certain biological processes including translocation of various substrates across membranes and non-transport-related processes such as translation of RNA and DNA repair.

Atypical choroid plexus papilloma (ACPP) A type of papilloma of the choroid plexus with atypical symptoms.

Atypical teratoid/rhabdoid tumors (AT/RT) A rare tumor with a high malignancy usually diagnosed in childhood. Although usually a brain tumor, AT/RT can occur anywhere in the central nervous system (CNS) including the spinal cord.

B cells A type of lymphocyte in the humoral immunity of the adaptive immune system. B cells can be distinguished from other lymphocytes, such as T cells and natural killer cells (NK cells), by the presence of a protein on the B cell's outer surface known as a B cell receptor (BCR).

Blood–meninges barrier (BLMB) The barrier of meninges that closes and protects the vessels that supply the brain and contains CSF between the pia mater and arachnoid maters.

Blood–brain barrier a selective mechanism opposing the passage of most ions and high molecular weight compounds from the blood to brain tissue located in a continuous layer of endothelial cells connected by tight junctions. Composed mainly of astrocytes.

Blood–choroid-plexus barrier (BCPB) Along with the BBB, a barrier that protects and provides nutrition to the brain. Made up by the choroid plexus, which separates the blood stream from the cerebrospinal fluid that circulates through the ventricles, the subarachnoidal spaces and along perivascular drainage pathways alongside brain vessels.

Blood–CSF barrier Stands for blood–cerebrospinal fluid barrier (CSF). A barrier located at the tight junctions that surround and connect the cuboidal epithelial cells on the surface of the choroid plexus.

Blood–leptomeningeal barrier (BLMB) The barrier of leptomeningeal cells that forms between the blood and the cerebrospinal fluid.

BMP signaling Stands for bone morphogenetic protein signaling. The signaling of a family of intracellular glycoproteins to induce new bone formation. These proteins influence bone remodeling, fracture healing, bone graft, integration, and heterotopic calcification.

Bone marrow The flexible tissue in the interior of bones.

Brain-derived neurotrophic factor (BDNF) A protein that, in humans, is encoded by the *BDNF* gene. It acts on certain neurons of the central nervous system (CNS) and the peripheral nervous system, helping to support the survival of existing neurons, and encourage the growth and differentiation of new neurons and synapses.

Brain metastases A cancer that has metastasized, or spread, to the brain from another location in the body.

Brain tumor What occurs when abnormal cells form within the brain. Can be malignant/cancerous, or can be benign.

Cancer General term frequently used to indicate any of various types of malignant neoplasms, most of which invade surrounding tissues, may metastasize to several sites, and are likely to recur after attempted removal and to kill the patient unless adequately treated.

CCL1 (MCP-1) A small glycoprotein secreted by activated T cells that belongs to a family of inflammatory cytokines known as chemokines.

CD4+ T cells T helper cells that express the protein CD4. are generally treated as having a predefined role as helper T cells within the immune system

CD68 Cluster of differentiation 68 is a glycoprotein, which binds to low-density lipoprotein. Its presence in macrophages also makes it useful in diagnosing conditions related to proliferation or abnormality of these cells, such as malignant histiocytosis, histiocytic lymphoma, and Gaucher's disease.

Central nervous system (CNS) The main structures involved in the nervous pathways, consisting of the brain and spinal cord.

Cerebral aqueduct Within the mesencephalon (or midbrain), contains cerebrospinal fluid, and connects the third ventricle in the diencephalon to the fourth ventricle within the region of the mesencephalon and metencephalon, located dorsal to the pons and ventral to the cerebellum.

Cerebral blood volume (rCBV) Blood supply to the brain in a given time. In an adult, CBF is typically **750 mL/min** or 15% of the cardiac output. This equates to an average perfusion of **50–54 mL** of blood per 100 g of brain tissue per minute.

Cerebrospinal fluid A fluid largely secreted by the choroid plexuses of the ventricles of the brain, filling the ventricles and the subarachnoid cavities of the brain and spinal cord.

Chemotherapy A category of cancer treatment that uses chemical substances, especially one or more anticancer drugs (chemotherapeutic agents) that are given as part of a standardized chemotherapy regimen.

Choline acetyltransferase (ChAT) A transferase enzyme responsible for the synthesis of the neurotransmitter acetylcholine.

Choroid epithelium The layer of cells that surround a core of capillaries and loose connective tissue that form the choroid plexus.

Choroid plexectomy Surgical removal of the choroid plexus.

Choroid plexus carcinoma A malignant neoplasm arising from the choroid plexus.

Choroid plexus metastasis Cancer that has spread to the choroid plexus from another part of the body.

Choroid plexus papilloma (CPP) A papilloma of the choroid plexus that begins in the ventricles and blocks the flow of CSF.

Choroid plexus papilloma An uncommon, benign neuroepithelial intraventricular tumor that can occur in the pediatric (more common) and adult population.

Choroid plexus tumor A rare type of cancer that occurs in the choroid plexus of the brain.

Choroid plexus A vascular proliferation or fringe of the tela choroidea in the third, fourth, and lateral cerebral ventricles; it secretes cerebrospinal fluid, thereby regulating to some degree the intraventricular pressure.

Choroidectomy Surgical procedure involving the removal of parts of the eye (middle layer of the eye between the retina and sclera). This is a treatment for intraocular melanoma.

Circulating tumor cells (CTCs) Cells that have shed into the vasculature from a primary tumor and circulate in the bloodstream. CTCs thus constitute seeds for subsequent growth of additional tumors (metastasis) in vital distant organs, triggering a mechanism that is responsible for the vast majority of cancer-related deaths.

Circumventricular organs Structures in the brain that are characterized by their extensive vasculature and lack of a normal blood–brain barrier.

Collagens The main structural protein of the various connective tissues in animals, mostly found in fibrous tissues.

Computed tomography (CT) scans Any computer-aided tomographic process, usually X-ray computed tomography, like its medical imaging counterparts uses irradiation (usually with X-rays) to produce three-dimensional representations of the scanned object externally and internally.

Craniospinal axis radiotherapy (CSI) Treatment option in patients with recurrent leukemia involving the central nervous system (CNS).

Crizotinib An anticancer drug approved for the treatment of some non-small-cell lung carcinoma (NSCLC) in the United States and some other countries.

Cryptococcus neoformans An encapsulated yeast that can live in both plants and animals. It can be associated with infection as well, including fungal meningitis and encephalitis.

CSF fluxes The change in flow of cerebrospinal fluid.

Cyclooxygenase 2 (COX2) An enzyme involved in the conversion of arachidonic acid to prostaglandin H2, an important precursor of prostacyclin and thromboxane A2. It is upregulated in cancer.

Cytomegalovirus A virus contracted by spreading of body fluids. It stays with the infected individual for life; however, it rarely causes symptoms unless they have a weak immune system or are pregnant.

Dendritic cells (DC) Antigen-presenting cells (also known as accessory cells) of the mammalian immune system. Their main function is to process antigen material and present it on the cell surface to the T cells of the immune system.

Desmosomes A type of junctional complex, they are localized spot-like adhesions randomly arranged on the lateral sides of plasma membranes.

Drug delivery Approaches, formulations, technologies, and systems for transporting a pharmaceutical compound in the body as needed to safely achieve its desired therapeutic effect.

E-selectins A cell adhesion molecule expressed only on endothelial cells activated by cytokines. Like other selectins, it plays an important part in inflammation.

EAE Experimental autoimmune encephalomyelitis. An animal model of brain inflammation. It is an inflammatory demyelinating disease of the central nervous system (CNS). It is mostly used with rodents and is widely studied as an animal model of the human CNS demyelinating diseases.

Embryonic forebrain The rostral-most (forward-most) portion of the brain in a developing embryo. Eventually it separates into the cerebrum and the diencephalon.

Encapsulated cell biodelivery A method of delivering something like a nerve growth factor to a patient with certain neurodegenerative diseases.

Endoscopic third ventriculostomy A surgical procedure for the treatment of hydrocephalus in which an opening is created in the floor of the third ventricle using an endoscope placed within the ventricular system through a burr hole. This allows the cerebrospinal fluid to flow directly to the basal cisterns, thereby shortcutting any obstruction.

Endothelial cell One of the simple squamous cells forming the lining of blood and lymph vessels and the inner layer of the endocardium.

Endothelium A layer of flat cells lining, especially blood and lymphatic vessels and the heart.

Ependymoma (papillary variant) A tumor that arises from the ependyma, a tissue of the central nervous system (CNS). The papillary variant shows similar characteristics with choroid plexus papilloma.

Epidermal growth factor receptor ligand A protein in the EGF family of proteins that has been shown to play a role in wound healing, cardiac hypertrophy, and heart development and function.

Epiplexus cells Phagocytes that may resemble monocytic cells or macrophageslie. They lie close to the ventricular surface of the choroid epithelium and apart from their primary function as scavenger cells, are engaged also in immunological responses and iron regulation in the ventricular system or the brain as a whole.

Epithelium The purely cellular avascular layer covering all free surfaces, cutaneous, mucous, and serous, including the glands and other structures derived therefrom.

Exosome cell-derived vesicles that are present in many and perhaps all biological fluids, including blood, urine, and cultured medium of cell cultures.

Extracellular matrix (ECM) A collection of extracellular molecules secreted by cells that provides structural and biochemical support to the surrounding cells.

Extracellular space (ECS) The part of a multicellular organism outside the cells proper, usually taken to be outside the plasma membranes and occupied by fluid.

Extravasation (cancer) Cancer cells exiting the capillaries and entering organs.

Fenestrated endothelium An endothelium that contains a small pore to allow rapid exchange of molecules between sinusoid blood vessels and surrounding tissue.

FGF-2 Fibroblast growth factor-2 (FGF2) is a wide-spectrum mitogenic, angiogenic, and neurotrophic factor that is expressed at low levels in many tissues and cell types and reaches high concentrations in the brain and pituitary. FGF2 has been implicated in a multitude of physiologic and pathologic processes, including limb development, angiogenesis, wound healing, and tumor growth.

Fibroblast growth factor-8 (FGF8) A protein of the FGF family important and necessary for setting up and maintaining the midbrain/hindbrain border, which plays the vital role of "organizer" in development, like the Spemann "organizer" of the gastrulating embryo.

Floor plate Floor plate is integral to the developing nervous system of vertebrate organisms. It is a specialized glial structure that spans the anteroposterior axis from the midbrain to the tail regions.

Foramen of Luschka An opening in each lateral extremity of the lateral recess of the fourth ventricle of the human brain, which also has a single median aperture. The two lateral apertures provide a conduit for cerebrospinal fluid to flow from the brain's ventricular system into the subarachnoid space.

Foramen of Magendie The median aperture (foramen of Magendie) drains cerebrospinal fluid from the fourth ventricle into the cisterna magna.

Foramina of Monro Channels that connect the paired lateral ventricles with the third ventricle at the midline of the brain. As channels, they allow cerebrospinal fluid produced in the lateral ventricles to reach the third ventricle and then the rest of the brain's ventricular system.

Fourth ventricle The fourth ventricle is one of the four connected fluid-filled cavities within the human brain. These cavities, known collectively as the ventricular system, consist of the left and right lateral ventricles, the third ventricle, and the fourth ventricle.

Fxyd1 A plasma membrane substrate for several kinases, including protein kinase A, protein kinase C, NIMA kinase, and myotonic dystrophy kinase. It is thought to form an ion channel or regulate ion channel activity.

Globulin The globulins are a family of globular proteins that have higher molecular weights than albumins and are insoluble in pure water but soluble in dilute salt solutions. Some globulins are produced in the liver, while others are made by the immune system.

Glucagon-like peptide-1 (GLP-1) An incretin that performs many physiological functions including increasing insulin secretion from the pancreas in a glucose-dependent manner, and decreases glucagon secretion from the pancreas by engagement of a specific G protein-coupled receptor.

Glymphatic system A functional waste clearance pathway for the mammalian central nervous system (CNS).

Granulocyte-macrophage colony-stimulating factor (GM-CSF) A monomeric glycoprotein secreted by macrophages, T cells, mast cells, NK cells, endothelial cells, and fibroblasts that functions as a cytokine.

Gross total resection (GTR) After surgical removal of a tumor, the entire visible tumor has been removed.

Hensen's node Organizer for gastrulation in vertebrates.

Heparin sulfate a linear polysaccharide found in all animal tissues. It occurs as a proteoglycan (HSPG) in which two or three HS chains are attached in close proximity to cell surface or extracellular matrix proteins. It regulates a variety of biological processes.

Hepatocyte growth factor (HGF) A paracrine cellular growth, motility, and morphogenic factor. It is secreted by mesenchymal cells and targets and acts primarily upon epithelial cells and endothelial cells.

HLA-DR An MHC class II cell surface receptor encoded by the human leukocyte antigen complex on chromosome 6 region 6p21.31. The complex of HLA-DR and its ligand, a peptide of 9 amino acids in length or longer, constitutes a ligand for the T-cell receptor (TCR).

Hyaluronic acid An anionic, nonsulfated glycosaminoglycan distributed widely throughout connective, epithelial, and neural tissues. One of the chief components of the extracellular matrix, hyaluronan contributes significantly to cell proliferation and migration, and may also be involved in the progression of some malignant tumors.

Hydrocephalus Medical condition in which there is an abnormal accumulation of cerebrospinal fluid (CSF) in the brain.

ICAM-1 Same as CD54 (Cluster of differentiation 54), it is a protein that in humans is encoded by the *ICAM1* gene. This gene encodes a cell surface glycoprotein that is typically expressed on endothelial cells and cells of the immune system.

Infiltration The act of permeating or penetrating into a substance, cell, or tissue; said of gases, fluids, or matter held in solution.

Insulin-like growth factor (IGF) Also called somatomedin C, a protein that in humans is encoded by the *IGF1* gene. It plays an important role in childhood growth and continues to have anabolic effects in adults.

Insulin-like growth factor-I (IGF-1) A hormone similar in molecular structure to insulin. It plays an important role in childhood growth and continues to have anabolic effects in adults.

Insulin-like growth factor-II (IGF-2) Believed to be a major fetal growth factor in contrast to insulin-like growth factor 1, which is a major growth factor in adults.

Internal carotid artery (ICA) A major paired artery, one on each side of the head and neck in human anatomy that supplies the brain, while the external carotid nourishes other portions of the head, such as face, scalp, skull, and meninges.

Interstitial fluid The fluid in spaces between the tissue cells, constituting about 16% of the weight of the body; closely similar in composition to lymph.

Intraparenchymal implantation A method of cell implantation using intraparenchymal cells to treat and repair stroke-related damages.

Intrathecal injection Referring to an injection occurring in the anatomic space or potential space inside a sheath, most commonly the arachnoid membrane of the brain or spinal cord.

Intravasation (cancer) The invasion of cancer cells through the basal membrane into a blood or lymphatic vessel.

Intraventricular tumor Benign tumors or lesions found within the ventricles of the brain. These tumors may arise from a variety of cells in the region and often obstruct the flow of cerebrospinal fluid and cause a buildup of pressure in the skull.

ISF (Interstitial fluid) A solution that bathes and surrounds the tissue cells of multicellular animals. It is the main component of the extracellular fluid, which also includes plasma and transcellular fluid.

Junction adhesion molecules (JAMs) Members of an immunoglobulin subfamily expressed by leukocytes and platelets as well as by epithelial and endothelial cells, in which they localize to cell–cell contacts and are specifically enriched at tight junctions

Laminins Proteins of the ECM that are a major component of the basal lamina. The laminins are an important and biologically active part of the basal lamina, influencing cell differentiation, migration, and adhesion, as well as phenotype and survival.

Lateral ventricle formation The lateral ventricle is developed from the central canal of the neural tube, specifically from the portion of the tube that is present in the developing prosenchephalon, and subsequently in the developing telencephalon.

Lateral ventricle A cavity shaped somewhat like a horseshoe in conformity with the general shape of the cerebral hemisphere. Part of the ventricular system in the brain.

Leptomeningeal metastases Refers to the spread of malignant cells through the CSF space. These cells originated in primary CNS tumors as well as from distant tumors that have metastasized.

Li–Fraumeni Syndrome A rare cancer-predisposed hereditary disorder characterized as autosomal dominant.

Lipopolysaccharide (LPS) Also called endotoxins, they are large molecules consisting of a lipid and a polysaccharide composed of O-antigen, outer core, and inner core joined by a covalent bond; they are found in the outer membrane of Gram-negative bacteria, and elicit strong immune responses in animals.

Low-density lipoprotein (LDL) A lipoprotein sometimes referred to as bad cholesterol because they can transport their content of fat molecules into artery walls, attract macrophages, and thus drive atherosclerosis.

Low-grade papillomas Human papillomavirus test result that shows early changes to the cells of the cervix, possibly indicating an HPV infection.

Lumbar puncture A diagnostic and at times therapeutic medical procedure. Diagnostically it is used to collect cerebrospinal fluid (CSF) to confirm or exclude conditions such as meningitis and subarachnoid hemorrhage and it may be used in the diagnosis of other conditions.

Lymphocytes A white blood cell formed in bone marrow and distributed throughout the body in lymphatic tissue (lymph nodes, spleen, thymus, tonsils, and Peyer patches), where it undergoes proliferation.

M2 monocyte-derived macrophages Macrophages that function in constructive processes like wound healing and tissue repair, and those that turn off damaging immune system activation by producing anti-inflammatory cytokines

Macrometastases Defined as a relatively large metastasis (see metastasis definition).

Macrophages A type of white blood cell that engulfs and digests cellular debris, foreign substances, microbes, cancer cells, and anything else that does not have the types of proteins specific to the surface of healthy body cells on its surface in a process called phagocytosis.

MadCAM-1 Addressins (other name) are proteins that are the ligands to the homing receptors of lymphocytes. The task of these ligands and their receptors is to determine which tissue the lymphocyte will enter next.

Magnetic resonance imaging (MRI) A medical imaging technique used in radiology to investigate the anatomy and physiology of the body in both health and disease. MRI scanners use magnetic fields and radio waves to form images of the body.

Mannitol A drug that treats early kidney failure by increasing urination. This helps the body get rid of extra fluids. Treats brain swelling and increased pressure in the eye.

Matrix metalloproteinases (MMP) Enzymes that are capable of degrading all kinds of extracellular matrix proteins.

Meningioma (papillary variant) A diverse set of tumors arising from the meninges, the membranous layers surrounding the central nervous system (CNS). The papillary variants are malignant, frequently show bone and parenchymatous invasion and have the potential for extracranial metastasis.

Mesencephalon See **Midbrain**

Metencephalon A developmental categorization of portions of the central nervous system (CNS) that is composed of the cerebellum and the pons.

Metrizamide A nonionic, water-soluble, iodinated radiographic contrast medium used in myelography and cisternography.

MHC class II receptor A family of molecules normally found only on antigen-presenting cells such as dendritic cells, mononuclear phagocytes, some endothelial cells, thymic epithelial cells, and B cells.

Microcolonization Achievement of more or less stable association of a microorganism with a given environment or microenvironment.

Midbrain A portion of the central nervous system (CNS) associated with vision, hearing, motor control, sleep/wake, arousal (alertness), and temperature regulation.

Middle cerebral artery One of the two large terminal branches (with anterior cerebral artery) of the internal carotid artery; it passes laterally around the pole of the temporal lobe, then posteriorly in the depth of the lateral cerebral fissure.

miRNA A small noncoding RNA molecule (containing about 22 nucleotides) found in plants, animals, and some viruses, which functions in RNA silencing and post-transcriptional regulation of gene expression.

Multidrug resistance proteins (MDR) A subfamily of ABC-binding cassette that contribute to drug resistance in cancer cells.

Multiple sclerosis Common demyelinating disorder of the central nervous system (CNS), causing patches of sclerosis (plaques) in the brain and spinal cord.

Myelencephalon Most posterior region of the embryonic hindbrain, from which the medulla oblongata develops.

N-acetylaspartate (NAA) The second-most concentrated molecule in the brain that performs many different functions, including being involved in fluid balance in the brain.

N-cadherin Calcium-dependent cell–cell adhesion glycoprotein that functions during gastrulation and is required for establishment of left–right asymmetry.

Nerve growth factor (NGF) A growth factor that can be used to prevent or reverse peripheral neuropathy. It is a small secreted protein that is important for the growth, maintenance, and survival of certain target neurons (nerve cells).

Neural plate The neuroectodermal region of the early embryo's dorsal surface that in later development is transformed into the neural tube and neural crest.

Neural tube The epithelial tube formed from the neuroectoderm of the early embryo by the closure of the neural groove; by complex processes of cell proliferation and organization, the neural tube develops into the spinal cord and brain.

Neuroinflammation Inflammation of the nervous tissue that may be initiated in response to a variety of cues, including infection, traumatic brain injury, or autoimmunity.

Neuroprotection The relative preservation of neuronal structure and/or function.

Neurotrophins A family of proteins that induce the survival, development, and function of neurons.

Neurulation The folding process in vertebrate embryos, which includes the transformation of the neural plate into the neural tube. The embryo at this stage is termed the neurula.

Nocardiosis An infectious disease affecting either the lungs (pulmonary nocardiosis) or the whole body. It is due to infection by bacterium of the genus Nocardia, most commonly *Nocardia asteroides* or *Nocardia brasiliensis*.

Organic anion transporters (Oats) A membrane transport protein or "transporter" that mediates the transport of mainly organic anions across the cell membrane.

Organic anion transporting polypeptides 1, Oatp2 An uptake transporter that plays an important role in drug disposition and is responsible for the hepatic uptake of drugs and endogenous compounds.

Organic anion transporting polypeptides 2, Oatp1 Similar to OATP2.

Organotropism The special affinity of chemical compounds or pathogenic agents for particular tissues or organs of the body.

Otx2 transcription factor A transcription factor that may play a role in brain and sensory organ development.

P-glycoprotein (P-gp) An important protein of the cell membrane that pumps many foreign substances out of cells. It is an ABC transporter.

P-selectin glycoprotein ligand 1 (PSGL-1) A high affinity counter-receptor for P-selectin on myeloid cells and stimulated T lymphocytes that plays a critical role in the tethering of these cells to activated platelets or endothelia expressing P-selectin.

P-selectins A cell adhesion molecule (CAM) on the surface of activated endothelial cells, which line the inner surface of blood vessels and activated platelets. They play an essential role in the initial recruitment of leukocytes (white blood cells) to the site of injury during inflammation.

Paget's "seed and soil" hypothesis A hypothesis by Stephen Paget about the nature of tumor metastasis. The outcome of cancer metastasis depends on multiple interactions between selected metastatic cells and homeostatic mechanisms unique to some organ microenvironments.

Paracellular diapedisis Transendothelial migration of leukocytes between endothelial cells.

Paravascular route A route of solute exchange between the interstitial fluid of the brain parenchyma and the CSF via paravascular spaces.

Pericytes One of the slender, mesenchymal-like cells found in close association with the outside wall of postcapillary venules; it is relatively undifferentiated and may become a fibroblast, endothelial cell, or smooth muscle cell.

Phospho-tau$_{181}$ (p-tau) Phosphorylated tau proteins can result in the self-assembly of tangles of paired helical filaments and straight filaments, which are involved in the pathogenesis of Alzheimer's disease, fronto-temporal dementia, and other tauopathies.

PICA – posterior inferior cerebellar artery The largest branch of the vertebral artery, is one of the three main arterial blood supplies for the cerebellum, part of the brain.

Pinocytotic vesicles A vesicle, a fraction of a micrometer in diameter, containing fluid or solute being ingested into a cell by endocytosis.

Positron emission tomography (PET) A nuclear medicine, functional imaging technique that produces a three-dimensional image of functional processes in the body by using a form of radioactive glucose as a tracer.

Posterior choroidal arteries (PChA) One of a pair of blood vessels that supply oxygenated blood to the posterior aspect of the brain (occipital lobe) in human anatomy.

Posterior choroidal artery Usually seen as two branches of the P2 segment of the posterior cerebral artery that supply the choroid plexus of the third ventricle (posterior medial choroidal artery) and parts of the choroid plexus of the lateral ventricle (posterior lateral choroidal artery).

Posterior communicating arteries (PCoA) Arteries at the base of the brain that form part of the circle of Willis and connect the ICA and PChA.

Posterior inferior cerebellar artery (PICA) The largest branch of the vertebral artery, is one of the three main arterial blood supplies for the cerebellum, part of the brain.

Prosencephalon The rostral-most (forward-most) portion of the brain that controls body temperature, reproductive functions, eating, sleeping, and any display of emotions.

Proteoglycans Heavily glycosylated proteins that are a major component of the animal extracellular matrix, the "filler" substance existing between cells in an organism.

Radiotherapy Therapy using ionizing radiation, generally as part of cancer treatment to control or kill malignant cells.

Receptor for advanced glycation end products (RAGE) A transmembrane receptor of the immunoglobulin super family hypothesized to have a causative effect in a range of inflammatory diseases such as diabetic complications, Alzheimer's disease, and even some tumors.

Receptor-related proteins (LRPs) Lipoprotein receptor proteins that mediate endocytosis of cholesterol-rich LDL.

Rho-associated kinases (ROCKs) Kinase involved mainly in regulating the shape and movement of cells by acting on the cytoskeleton.

Rhodamine 123 A tracer dye within water to determine the rate and direction of flow and transport.

Roof plate A neural structure in the embryonic nervous system, part of the dorsal side of the neural tube, that involves the communication of general somatic and general visceral sensory impulses. The caudal part later becomes the sensory axon part of the spinal cord.

Sjögrens syndrome A chronic autoimmune disease in which the body's white blood cells destroy the exocrine glands, specifically the salivary and lacrimal glands, that produce saliva and tears, respectively.

Sonic hedgehog signaling A signaling pathway that transmits information to embryonic cells required for proper development.

Spemann and Mangold organizer A cluster of cells in the developing embryo of an amphibian that induces development of the central nervous system (CNS).

ST6GALNAC5 An enzyme that belongs to a family of sialyltransferases that modify proteins and ceramides on the cell surface to alter cell–cell or cell–extracellular matrix interactions. It is a mediator of cancer cell passage through the blood–brain barrier.

Stereotactic radiosurgery A minimally invasive form of surgical intervention that makes use of a three-dimensional coordinate system to locate small targets inside the body to then perform destruction of precisely selected areas of tissue using ionizing radiation.

Sturge–Weber syndrome A rare congenital neurological and skin disorder. It is characterized by abnormal blood vessels on the brain surface. Normally, only one side of the brain is affected.

Subarchnoidal space The anatomic space between the arachnoid membrane and pia mater.

Subtotal resection (STR) A portion of the tumor can still be seen on a postoperative brain scan.

Subventricular zone A paired brain structure situated throughout the lateral walls of the lateral ventricles thought to be one of the two places where neurogenesis has been found to occur in the adult brain.

Superior cerebellar artery (SCA) An artery in the brain that arises near the termination of the basilar artery and supplies half of the cerebellum and parts of the midbrain.

T cells A type of lymphocyte (in turn, a type of white blood cell) that plays a central role in cell-mediated immunity.

T1- and T2-weighted images MRI images used to differentiate anatomical structures mainly on the basis of T1 or T2 values.

Tanycytes Special ependymal cells found in the third ventricle of the brain, and on the floor of the fourth ventricle and have processes extending deep into the hypothalamus. It is possible that their function is to transfer chemical signals from the cerebrospinal fluid to the central nervous system (CNS).

Tela choroidea A structure found in the ventricles of the brain (third and fourth).

TGF-β2 Transforming growth factor-beta 2 (TGF-β2) is a secreted protein known as a cytokine that performs many cellular functions and has a vital role during embryonic development. It is known to suppress the effects of interleukin-dependent T-cell tumors.

T_h1 cells The host immunity effectors against intracellular bacteria and protozoa.

T_h17 A subset of T helper cells producing interleukin 17 (IL-17) whose role is to provide antimicrobial immunity at epithelial/mucosal barriers.

T_h2 cells The host immunity effectors against extracellular parasites including helminths.

The metastatic cascade The multistep process of a tumor leaving its original location and spreading to another part of the body, or metastasizing.

Third ventricle One of four connected fluid-filled cavities comprising the ventricular system within the human brain. It is a median cleft in the diencephalon between the two thalami, and is filled with cerebrospinal fluid (CSF).

Thyroid hormone (T4) Tyrosine-based hormone produced by the thyroid gland that are primarily responsible for regulation of metabolism.

TNF-α Cell signaling protein (cytokine) involved in systemic inflammation and is one of the cytokines that make up the acute phase reaction.

Total-tau (T-tau) All tau protein isoforms, a measure of which can be used as a marker for neuro-degeneration. Tau is a microtubule-associated protein.

Toxoplasmosis A parasitic disease caused by the protozoan *Toxoplasma gondii*. The parasite infects most genera of warm-blooded animals, including humans, but the primary host is the felid (cat) family.

TP53 tumor suppressor A protein that conserves stability by preventing genome mutation and thus prevents, or suppresses cancer.

Transcellular diapedisis Transendothelial migration of leukocytes through individual endothelial cells.

Transendothelial migration The movement of leukocytes from different endothelial cells as needed, mainly during inflammation.

Transporters Neurotransmitter transporters are a class of membrane transport proteins that span the cellular membranes of neurons. Their primary function is to carry neurotransmitters across these membranes and to direct their further transport to specific intracellular locations.

Transthyretin (TTR) A serum and cerebrospinal fluid carrier of the thyroid hormone thyroxine (T4) and retinol-binding protein bound to retinol.

Trastuzumab (Herceptin) A monoclonal antibody that interferes with the *HER2/neu* receptor. Its main use is to treat certain breast cancers.

Tuberculosis A widespread, and in many cases fatal, infectious disease caused by various strains of mycobacteria, usually *Mycobacterium tuberculosis*.

Tumor necrosis factor (TNF) A group of cytokines that can cause cell death and have been implicated in cancer regression.

TWIST1 A basic helix–loop–helix transcription factor that has been implicated in cell lineage determination and differentiation.

VCAM-1 Vascular cell adhesion protein 1 also known as vascular cell adhesion molecule 1 (VCAM-1) or cluster of differentiation 106 (CD106) is a protein that in humans is encoded by the *VCAM1* gene. VCAM-1 functions as a cell adhesion molecule.

VEGF A signal protein produced by cells that stimulates vasculogenesis and angiogenesis. It is part of the system that restores the oxygen supply to tissues when blood circulation is inadequate.

Ventricular system A set of four interconnected cavities (ventricles) in the brain, where the cerebrospinal fluid is produced.

Ventriculo-cisternal perfusion ("Pappenheimer") Method for measuring cerebrospinal fluid formation rate.

Ventriculo-peritoneal shunt Surgery to treat excess cerebrospinal fluid in the brain.

Villous hypertrophy A condition characterized by overproduction of cerebrospinal fluid by bilaterally symmetric and enlarged, yet histologically normal, choroid plexis.

Vimentin positivity Vimentin is a type of protein whose presence, or positivity, turns out to be an indicator for sarcomas and other mesenchymal neoplasm.

Virchow–Robin spaces The immunological spaces between the arteries and veins (not capillaries) and pia mater that can be expanded by leukocytes.

Whole brain radiation therapy (WBRT) A mainstay of treatment in patients with both identifiable brain metastases and prophylaxis for microscopic disease.

Wnt signaling A group of signal transduction pathways made of proteins that pass signals from outside of a cell through cell surface receptors to the inside of the cell.

YKL-40 A secreted glycoprotein whose exact physiological role is not known, but it has been implicated in the development of inflammatory diseases. YKL-40 is secreted by various cell types including macrophages, chondrocytes, and some types of cancer cells.

Zonula occludens Tight junctions (other name) are closely associated areas of two cells whose membranes join together forming a virtually impermeable barrier to fluid.

Subject Index